AF662554

A Modern Book On Environmental Sociology

A Modern Book On Environmental Sociology

H K Pathak

CYBER TECH PUBLICATIONS
4264/3, Ansari Road, Daryaganj N.Delhi - 110002
Phone : 011-43559448, Fax : 011-23280028
Email : cybertechpublications@gmail.com

CYBER TECH PUBLICATIONS
4264/3, Ansari Road, Daryaganj
New Delhi - 110002 (India)
Phone : 011-23244078, 011-43559448,
Fax : 011-23280028
E-mail : cybertechpublications@gmail.com
Website : www.cybertechpublications.com

A Modern Book On Environmental Sociology

First Published : 2010

ISBN 978-81-7884-606-4

Price,800.00

© Reserved

[No part of this publication may be reproduced, stored in a retrieval system or transmitted, in any form or by any means, mechanical, photocopying, recording or otherwise, without prior written permission of the publisher].

Printed at :

New Green Print Solutions, New Delhi 110078

CONTENTS

Preface

> "Make no man your friend before inquiring how he has used his former friends; for you must expect his to treat you as he treated them. Be slow to give your friendship, but when you have given it, strive to make it lasting; for it is as reprehensible to make many changes in one's associates as to have no friends at all. Neither test your friends to your own injury nor be willing to forego a test of your companions."
>
> —*Isocrates*

It is significant to note that children's rights are the human rights of children with particular attention to the rights of special protection and care afforded to the young, including their right to association with both biological parents, human identity as well as the basic needs for food, universal state-paid education, health care and criminal laws appropriate for the age and development of the child.

Interpretations of children's rights range from allowing children the capacity for autonomous action to the enforcement of children being physically, mentally and emotionally free from abuse, though what constitutes "abuse" is a matter of debate. Other definitions include the rights to care and nurturing.

As the famous adage goes on 'child is the father of man', it's the duty of everyone to look after the rights of child. Amidst plenty of literature, this book compiles valuable information on rights of child. The vital information is gathered under the following chapters: Understanding Child and

Childhood; Focus on Children's Rights; Towards Holistic Development of Children; National Plan of Action on Children; Issues in Child Protection Work; UN Action and National Policies for Children in Developing Countries; Rights of the Child and NGOs; Towards a Society for All; UN Convention on the Rights of the Child; Appendices and Bibliography etc.

Definitely, academics, policy planners and child and human rights activists will find the information of utmost use.

As is obvious, such works cannot be accomplished single-handedly. I have consulted a number of authoritative works, penned by various experts in the field. Besides printed works, websites are also profusely used and consulted. I express my deep sense of gratitude to all those scholars and their works. My seniors, colleagues, friends and family members have helped and cooperated me while working on this project. I am thankful to them. Last but not the least, I am thankful to my publishers for evincing keen interest in this work and materializing it elegantly in the present shape.

Comments from all quarters are welcome to improve the quality of contents.

—David Malcolm

1

Understanding Child and Childhood

> "The older I grow, the more I appreciate children. Now, as I near my birthday, I salute them again. Children are the most wholesome part of the race, the sweetest, for they are freshest from the Hand of God. Whimsical, ingenious, mischievous, they fill the earth with joy and good humor. We adults live a life of apprehension as to what they will think of us; a life of defense against their terrifying energy; a life of hard work to live up to their great expectations. We put them to bed with a sense of relief—and greet them in the morning with delight and anticipation. We envy them the freshness of adventure and the discovery of life. In all these ways, children add to the wonder of being alive. In all these ways, they help to keep us young."
>
> —*Herbert Hoover*

As we intend to study rights of child in this volume, it is imperative to have a thorough understanding of child and childhood.

It is interesting to note that biologically, a child is generally a human between the stages of birth and puberty. Some vernacular definitions of a child include the fetus, as being an unborn child. The legal definition of "child" generally refers to a minor, otherwise known as a person younger than the age of majority. "Child" may also describe a relationship with a parent (such as sons and daughters of any age) or, metaphorically, an authority figure, or signify group membership in a clan, tribe, or religion; it can also signify being strongly affected by a specific time, place, or circumstance, as in "a child of nature" or "a child of the Sixties".

In fact, United Nations Convention on the Rights of the Child defines a child as "a human being below the age of 18 years unless under the law applicable to the child, majority is attained earlier". Ratified by 192 of 194 member countries. Some English definitions of the word 'child' include the fetus and the unborn. Biologically, a child is anyone between birth and puberty or in the developmental stage of childhood, between infancy and adulthood. Children generally have fewer rights than adults and are classed as not able to make serious decisions, and legally must always be under the care of a responsible adult.

Recognition of childhood as a state different from adulthood began to emerge in the 16th and 17th centuries. Society began to relate to the child not as a miniature adult but as a person of a lower level of maturity needing adult protection, love and nurturing. This change can be traced in painting: In the Middle Ages, children were portrayed in art as miniature adults with no childish characteristics. In the 16th century, images of children began to acquire a distinct childish appearance. From the late 17th century onwards, children were shown playing.

Toys and literature for children also began to develop at this time. Social attitudes toward children differ around the world in various cultures. These attitudes have changed over time. A 1988 study on European attitudes toward the centrality of children found that Italy was more child-centric and Holland less child-centric, with other countries, such as Austria, Great Britain, Ireland and West Germany falling in between.

Socialization

All children go through stages of social development. An infant or very young child will play alone happily. If another child wanders onto the scene, he or she may be physically attacked or pushed out of the way. Next, the child is able to play with another child, gradually learning to share and take turns. Eventually the group grows larger, to three or four children.

By the time a child enters kindergarten, he or she is usually able to join in and enjoy group experiences. Children with ADHD and learning disabilities may need extra help in developing social skills. The impulsive characteristics of an ADHD child may lead to poor peer relationships. Children with poor attention spans may not tune in to social cues in their environment, making it difficult for them to learn social skills through experience.

Age of responsibility

In fact, age at which children are considered responsible for their society-bound actions (e. g. marriage, voting, etc.) has also changed over time, and this is reflected in the way they are treated in courts of law. In Roman times, children were regarded as not culpable for crimes, a position later adopted by the Church. In the nineteenth century, children younger than seven years old were believed incapable of crime. Children from the age of seven forward were considered responsible for their actions. Therefore, they could face criminal charges, be sent to adult prison, and be punished like adults by whipping, branding or hanging. Today, in many countries like Canada and the United States, children twelve and older are held responsible for their actions and may be sent to special correctional institutions, such as juvenile hall.

Surveys have found that at least 25 countries around the world have no specified age for compulsory education. Minimum employment age and marriage age also vary. In at least 125 countries, children aged 7–15 may be taken to court and risk imprisonment for criminal acts. In some countries, children are legally obliged to go to school until they are 14 or 15 years old, but may also work before that age. A child's right to education is threatened by early marriage, child labour and imprisonment. During the early 1600s in England, life expectancy was only about 35 years, largely because two-thirds of all children died before the age of four. During the Industrial Revolution, the life expectancy of children increased dramatically.

According to population health experts, child mortality rates have fallen sharply since the 1990s. Deaths of children under the age of five are down by 42% in the United States, while Serbia and Malaysia have cut their rates by nearly 70%.

China's one-child policy forces some couples to have no more than one child. China's population policy has been credited with a very significant slowing of China's population growth which had been higher before the policy was implemented. It has come under criticism that the implementation of the policy has involved forced abortions and forced sterilization. However, while the punishment of "unplanned" pregnancy is a fine, both forced abortion and forced sterilization can be charged as intentional assault, which is punished with up to 10 years' imprisonment. If born with another child and kept, parents must pay a large fine for every day he/she is alive.

Now a look at childhood.

Childhood

In fact, childhood is the age span ranging from birth to adolescence. In developmental psychology, childhood is divided up into the developmental stages of toddlerhood (learning to walk), early childhood (play age), middle childhood (school age), and adolescence (puberty through post-puberty).

The term childhood is non-specific and can imply a varying range of years in human development.

Developmentally, it refers to the period between infancy and adulthood. In common terms, childhood is considered to start from birth. Some consider that childhood, as a concept of play and innocence, ends at adolescence. In many countries, there is an age of majority when childhood officially ends and a person legally becomes an adult. The age ranges anywhere from 15 to 21, with 18 being the most common.

Early Childhood

Early childhood follows the infancy stage, and begins with toddlerhood when the child begins speaking or taking steps independently. While toddlerhood ends around age three when the child becomes less dependent on parental assistance for basic needs, early childhood continues approximately through years seven or eight. According to the National Association for the Education of Young Children, early childhood spans the human life from birth to age eight.

Middle Childhood

Middle childhood begins at around age seven or eight, approximating primary school age and ends around puberty, which typically marks the beginning of adolescence.

Adolescence

Adolescence is usually determined by the onset of puberty. However, puberty may also begin in preadolescents. The end of adolescence and the beginning of adulthood varies by country and by function, and even within a single nation-state or culture there may be different ages at which an individual is considered to be (chronologically and the legally) mature enough to be entrusted by society with certain tasks. It has been argued that childhood is not a natural phenomenon but a creation of society. Philippe Ariès, an important French medievalist and historian, pointed this out in his book Centuries of Childhood.

This theme was then taken up by Cunningham in his book the Invention of Childhood (2006) which looks at the historical aspects of childhood from the Middle Ages to what he refers to as the Post War Period of the 1950s, 1960s and 1970s. Ariès published a study in 1961 of paintings, gravestones, furniture, and school records. He found that before the 17th-century, children were represented as mini-adults. Since then historians have increasingly researched childhood in past times. Before Ariès, George Boas had published The Cult of Childhood.

During the Renaissance, artistic depictions of children increased dramatically in Europe. This did not impact the social attitude to children much, however—see the article on child labour. The man usually credited with - or accused of - creating the modern notion of childhood is Jean Jacques Rousseau. Building on the ideas of John Locke and other 17th-century liberal thinkers, Rousseau formulated childhood as a brief period of sanctuary before people encounter the perils and hardships of adulthood. "Why rob these innocents of the joys which pass so quickly," Rousseau pleaded.

"Why fill with bitterness the fleeting early days of childhood, days which will no more return for them than for you?"

The Victorian Era has been described as a source of the modern institution of childhood. Ironically, the Industrial Revolution during this era led to an increase in child labour, but due to the campaigning of the Evangelicals, and efforts of author Charles Dickens and others, child labour was gradually reduced and halted in England via the Factory Acts of 1802-1878.

The Victorians concomitantly emphasized the role of the family and the sanctity of the child, and broadly speaking, this attitude has remained dominant in Western societies since then. In the contemporary era Joe L. Kincheloe and Shirley R. Steinberg have constructed a critical theory of childhood and childhood education that they have labeled kinderculture. Kincheloe and Steinberg make use of multiple research and theoretical discourses (the bricolage) to study childhood from diverse perspectives—historiography, ethnography, cognitive research, media studies, cultural studies, political economic analysis, hermeneutics, semiotics, content analysis, etc. Based on this multiperspectival inquiry, Kincheloe and Steinberg contend that new times have ushered in a new era of childhood. Evidence of this dramatic cultural change is omnipresent, but many individuals in the late 20th century and early 21st century have not yet noticed it.

When Steinberg and Kincheloe wrote the first edition of Kinderculture: The Corporate Construction of Childhood in 1997 (second edition, 2004) many people who made their living studying, teaching, or caring for children were not yet aware of the nature of the changes in childhood that they encountered daily. In the domains of psychology, education, and to a lesser degree sociology and cultural studies few observers before kinderculture had studied the ways that the information explosion so characteristic of our contemporary era (hyperreality) had operated to undermine traditional notions of childhood and change the terrain of childhood education.

Those who have shaped, directed and employed contemporary information technology have played an exaggerated role in the reformulation of childhood. Of course, information technology alone, Kincheloe and Steinberg maintain, has not produced a new era of childhood. Obviously, numerous social, cultural, and political economic factors have operated to produce such changes.

The central purpose of kinderculture is to socially, culturally, politically, and economically situate the changing historical status of childhood and to specifically interroge the ways diverse media have helped construct what Kincheloe and Steinberg call "the new childhood." Kinderculture understands that childhood is an ever-changing social and historical artifact—not simply a biological entity. Because many psychologists have argued that childhood is a natural phase of growing up, of becoming an adult, Kincheloe and Steinberg coming from an educational context saw kinderculture as a corrective to such a "psychologization" of childhood.

It is pertinent to note that the geographies of childhood involves how (adult) society perceives the very idea of childhood and the many ways the attitudes and behaviors of adults affects children's lives. This includes ideas about the surrounding environment of children and its related implications. This is similar in some respects to children's geographies which examines the places and spaces in which children live.

Modern Concepts of Childhood

The concept of childhood appears to evolve and change shape as lifestyles change and adult expectations alter. Some believe that children should not have any worries and should not have to work; life should be happy and trouble-free. Childhood is usually a mixture of happiness, wonder, angst and resilience. It is generally a time of playing, learning, socializing, exploring, and worrying in a world without much adult interference, aside from parents.

It is a time of learning about responsibilities without having to deal with adult responsibilities. Childhood is often retrospectively viewed as a time of innocence, which is generally viewed as a positive term, connoting an optimistic view of the world, in particular one where the lack of knowledge stems from a lack of wrongdoing, whereas greater knowledge comes from doing wrong. A "loss of innocence" is a common concept, and is often seen as an integral part of coming of age. It is usually thought of as an experience or period in a child's life that widens their awareness of evil, pain or the world around them. This theme is demonstrated in the novels To Kill a Mockingbird and Lord of the Flies. The fictional character Peter Pan was the embodiment of a childhood that never ends.

Nature Deficit Disorder

Nature Deficit Disorder, a term coined by Richard Louv in his 2005 book Last Child in the Woods, refers to the alleged trend in the United States that children are spending less time outdoors, resulting in a wide range of behavioral problems. With the advent of the computer, video games and television, children have more reasons to stay inside rather than outdoors exploring. "The average American child spends 44 hours a week with electronic media". Parents are also keeping children indoors in order to protect them from their growing fear of "stranger danger".

Recent research has drawn a further contrast between the declining number of National Park visits in the United States and increasing consumption of electronic media by children.

Play

Play is essential to the cognitive, physical, social, and emotional well-being of children. It offers children opportunities for physical (running, jumping, climbing, etc.), intellectual (social skills, community norms, ethics and general knowledge) and emotional development (empathy, compassion, and friendships). Unstructured play encourages creativity and imagination. Playing and interacting with other children, as well as some adults, provides opportunities for friendships, social interactions, conflicts and resolutions. It is through play that children at a very early age engage and interact in the world around them. Play allows children to create and explore a world they can master, conquering their fears while practicing adult roles, sometimes in conjunction with other children or adult caregivers. Undirected play allows children to learn how to work in groups, to share, to negotiate, to resolve conflicts, and to learn self-advocacy skills.

However, when play is controlled by adults, children acquiesce to adult rules and concerns and lose some of the benefits play offers them, particularly in developing creativity, leadership, and group skills. Play is considered to be so important to optimal child development that it has been recognized by the United Nations High Commission for Human Rights as a right of every child.

Children who are being raised in a hurried and pressured style may limit the protective benefits they would gain from child-driven play. The initiation of play in a classroom setting allows teachers and students to interact through playfulness associated with a learning experience. Therefore, playfulness aids the interactions between adults and children in a learning environment. "Playful Structure" means to combine informal learning with formal learning to produce an effective learning experience for children at a young age.

Street Culture

It is pertinent to note that children's street culture refers to the cumulative culture created by young children and is sometimes referred to as their secret world. It is most common in children between the ages of seven and twelve. It is strongest in urban working class industrial districts where children are traditionally free to play out in the streets for long periods without supervision. It is invented and largely sustained by children themselves with little adult interference. Young children's street culture usually takes place on quiet backstreets and pavements, and along routes that venture out into local parks, playgrounds, scrub and wasteland, and to local shops. It often imposes imaginative status on certain sections of the urban realm (local buildings, kerbs, street objects, etc.). Children designate specific areas that serve as informal meeting and relaxation places (see: Sobel, 2001).

An urban area that looks faceless or neglected to an adult may have deep 'spirit of place' meanings in to children. Since the advent of indoor distractions such as video games, and television, concerns have been expressed about the vitality - or even the survival - of children's street culture.

Childhood Studies

Childhood Studies (CS) is a multi-disciplinary field that utilizes all areas of study to understand childhoods as experienced by children today. CS draws on scholarship in the social sciences (specifically anthropology, economics, history, and sociology), the humanities (especially literature, religion, philosophy, and the fine arts), and the behavioral sciences (with an emphasis on psychology). Childhood Studies is a relatively new academic field. Discussions of its development can be found in Mary Jane Kehily's An Introduction to Childhood Studies and in Jean and Richard Mills' Childhood Studies: A Reader in Perspectives of Childhood. According to Kehily, dedicated degree programs in childhood studies are a recent development.

Gertrud Lenzer writes an article in August, 1991 titled "Is There Sufficient Interest to Establish a Sociology of Children?" in the journal Footnotes. James S. Coleman, ASA President-Elect writes as an addendum to this article, "I believe that the call for increased attention to the sociology of children is well-placed, and it may be that a new section of the ASA is the best way to do that."

Edward Rothstein writes in an article in The New York Times titled How Childhood Has Changed! (Adults Too), "Three years ago, at Brooklyn College, the sociologist Gertrud Lenzer founded a pioneering program in children studies applying anthropology, psychology, literature and history to the study of children and adolescence. Nearly 100 students now minor in the area; other schools have begun similar programs". In the Sunday Reader of The Dallas Morning News an article titled Lenzer Champions Growing Field of Children's Studies, Ira J. Hadnot writes that Lenzer created a novel program that is being copied across the country.

In an article in the Lion and the Unicorn in 2001 Lenzer provides a short history of Children's Studies. She writes the new and interdisciplinary field of Children's Studies was founded in 1991 at Brooklyn College of The City University of New York.

The Open University degree program was one of the first in the UK and was begun in 2003. In the United States there are now dozens of children/childhood "modules," minors, or concentrations within the degree programs of academic disciplines, as well as BA and Masters programs across the globe. York University in Canada quotes Lenzer in their proposal for a new degree program in Children Studies writing "In a special issue of The Lion and the Unicorn in 2001, Gertrud Lenzer, co-founder and Director of the Children's Studies Program and Children's Studies Center at Brooklyn College, provides a brief history of, and rationale for, the emerging field of children's studies. According to Lenzer, before the 1990s, most disciplines in the arts and sciences failed to "provide a special focus on children"; indeed, advertisers and politicians "discovered" childhood before scholars did. Only during the past two decades, Lenzer argues, has "an increasing number of disciplines in the arts and sciences . . . begun to manifest an interest in children and youth. In the humanities, these growing subfields include children's literature, the history of childhood, and the philosophy of children."

However, Lenzer suggests, even "the recent sharpening focus on children and youth in the humanities, social sciences, and international law" limited the efficacy of studies of children and childhood because "the intellectual division of labor in children-related scholarship across the disciplines was largely adding new subspecialties of and within the disciplines themselves." By contrast, Lenzer emphasizes the need for holistic, interdisciplinary—indeed, humanities-based— approaches to children's studies: ". . . [C]hildren are not fully characterized by psychological developmental processes, nor . . . by any single perspective. . . . [C]hildren also exist . . . as individuals, as a social and cultural class, and as a historical generation." York's Children's Studies program adopts many goals Lenzer proposes for this emerging area "as a genuinely interdisciplinary, multidisciplinary new field of study. By bringing carefully chosen knowledge of children from different studies to bear upon the class or category of children to students in a Liberal Arts course of learning, we hope that a more holistic understanding of children and childhood should emerge, which in the end will represent more than simply the sum of its parts. . . . Children's Studies . . . makes the ontological claim that children must be viewed in their fullness as human beings."

The importance of this field of study was underscored in March 2005, at the "Off to See the Wizard: Quests and Memory in Children's Literature" conference, when Roni Natov, author of Poetics of Childhood (2002), suggested that "interdisciplinary childhood studies" would transform future understandings of children and children's literature." Rutgers University, Camden, New Jersey, developed the first Childhood Studies Department in the United States to award degrees from BA through PhD. This is an multi-disciplinary department in which PhD students study a range of methodologies to explore cultural constructions of childhood. Rutgers University also operates The Center for Children's and Childhood Studies, the Rutgers University Press Book Series in Childhood Studies, and The Exploring Childhood Studies listserve, an online community of nearly 1300 academics and practitioners studying children and childhood in every discipline and around the world.

It is pertinent to note that evolution of Childhood Studies as an academic field has much in common with the development of other types of multi-disciplinary scholarship such as African-American Studies and Women's Studies. Each was initiated for the purpose of bringing to the academy points of view that had been underrepresented, if not repressed. New developments in the field of childhood studies include the founding of international childhood studies. This is interested in how global and international structures and processes shape children's lives and cultures of childhood. Birkbeck College offers Msc and PhD study in International

Childhood Studies in the Department of Geography, Environment and Development Studies (http://www.bbk.ac.uk/study/pg2011/childhoodstudies/TMSINTCH.html), reflecting the strong interest of this field in the intersections between childhood and international development. Karen Wells writes in Childhood in a Global Perspective (Polity 2009) that 'global processes and structures - especially the increasing influence of international law and international NGOs are reshaping childhood' (2009:1).

In the next chapter we will illustrate a vivid account of Rights of Child.

REFERENCES

Ariès, Philippe. Centuries of Childhood: A Social History of Family Life. New York: Alfred A. Knopf, 1962.

Boas, George. The Cult of Childhood. London: Warburg, 1966.

Brown, Marilyn R., ed. Picturing Children: Constructions of Childhood between Rousseau and Freud. Aldershot: Ashgate, 2002.

Buckingham, David. After the Death of Childhood: Growing Up in the Age of Electronic Media. Blackwell Publishers, 2000.

Bunge, Marcia J., ed. The Child in Christian Thought. Grand Rapids, MI: William B. Eerdmans Publishing Co., 2001.

Calvert, Karin. Children in the House: The Material Culture of Early Childhood, 1600-1900. Boston: Northeastern University Press, 1992.

Cannella, Gaile and Joe L. Kincheloe. "Kidworld: Childhood Studies, Global Perspectives, and Education". New York: Peter Lang, 2002.

Cleverley, John and D.C. Phillips. Visions of Childhood: Influential Models from Locke to Spock. New York: Teachers College, 1986.

Cunningham, Hugh. Children and Childhood in Western Society since 1500. London: Longman, 1995.

Cunnington, Phillis and Anne Buck. Children's Costume in England: 1300 to 1900. New York: Barnes & Noble, 1965.

deMause, Lloyde, ed. The History of Childhood. London: Souvenir Press, 1976.

Higonnet, Anne. Pictures of Innocence: The History and Crisis of Ideal Childhood. London: Thames and Hudson Ltd., 1998.

Immel, Andrea and Michael Witmore, eds. Childhood and Children's Books in Early Modern Europe, 1550-1800. New York: Routledge, 2006.

Kincaid, James R. *Child-Loving: The Erotic Child and Victorian Culture.* New York: Routledge, 1992.

Knörr, Jacqueline, ed. *Childhood and Migration. From Experience to Agency.* Bielefeld: Transcript, 2005.

Melchiorre, A. (2004) At What Age?...are school-children employed, married and taken to court?

Müller, Anja, ed. *Fashioning Childhood in the Eighteenth Century:* Age and Identity. Burlington, VT: Ashgate, 2006.

O'Malley, Andrew. The Making of the Modern Child: Children's Literature and Childhood in the Late Eighteenth Century. London: Routledge, 2003.

Pinchbeck, Ivy and Margaret Hewitt. Children in English Society. 2 vols. London: Routledge, 1969.

Pollock, Linda A. Forgotten Children: Parent-child relations from 1500 to 1900. Cambridge: Cambridge University Press, 1983.

Rachel K. Jones and April Brayfield, Life's greatest joy?: European attitudes toward the centrality of children. *Social Forces,* Vol. 75, No. 4, Jun 1997. 1,239-69 pp. Chapel Hill, North Carolina.

Sommerville, C. John. *The Discovery of Childhood in Puritan England.* Athens: University of Georgia Press, 1992.

Steinberg, Shirley R. and Joe L. Kincheloe. *Kinderculture: The Corporate Construction of Childhood.* Westview Press Inc., 2004.

Stone, Lawrence. The Family, Sex and Marriage in England 1500-1800. New York: Harper and Row, 1979.

W.J. Rorabaugh, Donald T. Critchlow, Paula C. Baker (2004). "America's promise: a concise history of the United States". Rowman & Littlefield. p.47.

Zornado, Joseph L. *Inventing the Child: Culture, Ideology, and the Story of Childhood.* New York: Garland, 2001.

2

Focus on Children's Rights

> "If a child is not disciplined and taught self-con-trol early in the home, the grown-up world will take care of him later on, perhaps cruelly and when it is too late. A child curbed, taught obedience, spanked when he is young, rarely requires punishment when he hits his teens."
>
> —*John Warren Hill*

It is significant to note that children's rights are the human rights of children with particular attention to the rights of special protection and care afforded to the young, including their right to association with both biological parents, human identity as well as the basic needs for food, universal state-paid education, health care and criminal laws appropriate for the age and development of the child.

Interpretations of children's rights range from allowing children the capacity for autonomous action to the enforcement of children being physically, mentally and emotionally free from abuse, though what constitutes "abuse" is a matter of debate. Other definitions include the rights to care and nurturing.

"A child is any human being below the age of eighteen years, unless under the law applicable to the child, majority is attained earlier."

According to Cornell University, a child is a person, not

a *subperson*, and the parent has absolute interest and possession of the child, but this is very much an American view. The term "child" does not necessarily mean minor but can include adult children as well as adult nondependent children. There are no definitions of other terms used to describe young people such as "adolescents", "teenagers," or "youth" in international law, but the children's rights movement is considered distinct from the youth rights movement. The field of children's rights spans the fields of law, politics, religion, and morality.

It must be remembered that as minors by law children do not have autonomy or the right to make decisions on their own for themselves in any known jurisdiction of the world. Instead their adult caregivers, including parents, social workers, teachers, youth workers and others, are vested with that authority, depending on the circumstances. Some believe that this state of affairs gives children insufficient control over their own lives and causes them to be vulnerable.

Louis Althusser has gone so far as describe this legal machinery, as it applies to children, as "repressive state apparatuses". Structures such as government policy have been held by some commentators to mask the ways adults abuse and exploit children, resulting in child poverty, lack of educational opportunities, and child labor.

On this view, children are to be regarded as a minority group towards whom society needs to reconsider the way it behaves. However, there is no evidence that such views are widely shared in society. Researchers have identified children as needing to be recognized as participants in society whose rights and responsibilities need to be recognized at all ages.

Consensus on defining children's rights has become clearer in the last fifty years. A 1973 publication by Hillary Clinton (then an attorney) stated that children's rights were a "slogan in need of a definition". According to some researchers, the notion of children's rights is still not well defined, with at least one proposing that there is no singularly accepted definition or theory of the rights held by children. Children's rights law is defined as the point where the law intersects with a child's life. That includes juvenile delinquency, due process for children involved in the criminal justice system, appropriate representation, and effective rehabilitative services; care and protection for children in state care; ensuring education for all children regardless of their origin, race, gender, disabilities, or abilities, and; health care and advocacy.

Types of Rights

Simultaneously, children's rights are defined in numerous ways, including a wide spectrum of civil, cultural, economic, social and political rights. Rights tend to be of two general types: those advocating for children as autonomous persons under the law and those placing a claim on society for protection from harms perpetrated on children because of their dependency. These have been labeled as the right of empowerment and as the right to protection.

Categories

- Provision: Children have the right to an adequate standard of living, health care, education and services, and to play and recreation. These include a balanced diet, a warm bed to sleep in, and access to schooling.
- Protection: Children have the right to protection from abuse, neglect, exploitation and discrimination. This includes the right to safe places for children to play; constructive child rearing behavior, and acknowledgment of the evolving capacities of children.
- Participation: Children have the right to participate in communities and have programs and services for themselves. This includes children's involvement in libraries and community programs, youth voice activities, and involving children as decision-makers.

In a similar fashion, the Child Rights Information Network, categorizes rights into two groups:

- Economic, social and cultural rights, related to the conditions necessary to meet basic human needs such as food, shelter, education, health care, and gainful employment. Included are rights to education, adequate housing, food, water, the highest attainable standard of health, the right to work and rights at work, as well as the cultural rights of minorities and indigenous peoples.
- Environmental, cultural and developmental rights, which are sometimes called "third generation rights," and including the right to live in safe and healthy environments and that groups of people have the right to cultural, political, and economic development.

Amnesty International openly advocates four particular children's rights, including the end to juvenile incarceration without parole, an end to the recruitment of military use of children, ending the death penalty for people under 21, and raising awareness of human rights in the classroom. Human Rights Watch, an international advocacy organization, includes child labor, juvenile justice, orphans and abandoned children, refugees, street children and corporal punishment.

Scholarly study generally focuses children's rights by identifying individual rights. The following rights "allow children to grow up healthy and free":

- Freedom of speech
- Freedom of thought
- Freedom from fear
- Freedom of choice and the right to make decisions
- Ownership over one's body.

Other issues affecting children's rights include the military use of children, sale of children, child prostitution and child pornography.

Difference Between Children's Rights and Youth Rights

"In the majority of jurisdictions, for instance, children are not allowed to vote, to marry, to buy alcohol, to have sex, or to engage in paid employment." Within the youth rights movement, it is believed that the key difference between *children's* rights and *youth* rights is that children's rights supporters generally advocate the establishment and enforcement of protection for children and youths, while youth rights (a far smaller movement) generally advocates the expansion of freedom for children and/or youths and of rights such as suffrage.

Parental Rights

Parents affect the lives of children in a unique way, and as such their role in children's rights has to be distinguished in a particular way. Particular issues in the child-parent relationship include child neglect, child abuse, freedom of choice, corporal punishment and child custody. There have been theories offered that provide parents with rights-based practices that resolve the tension between "commonsense parenting" and children's rights. The issue is particularly relevant in legal proceedings that affect the potential emancipation of minors, and in cases where children sue their parents.

A child's rights to a relationship with both their parents is increasingly recognized as an important factor for determining the best interests of the child in divorce and child custody proceedings. Some governments have enacted laws creating a rebuttable presumption that shared parenting is in the best interests of children.

It is interesting to note that 1796 publication of Thomas Spence's *Rights of Infants* is among the earliest English-language assertions of the rights of children. Throughout the 20th century children's rights activists organized for homeless children's rights and public education. The 1927 publication of *The Child's Right to Respect* by Janusz Korczak strengthened the literature surrounding the field, and today dozens of international organizations are working around the world to promote children's rights.

It is significant to note that opposition to children's rights far outdates any current trend in society, with recorded statements against the rights of children dating to the 13th century and earlier. Opponents to children's rights believe that young people need to be protected from the adultcentric world, including the decisions and responsibilities of that world. In the dominate adult society, childhood is idealized as a time of innocence, a time free of responsibility and conflict, and a time dominated by play.

The majority of opposition stems from concerns related to national sovereignty, states' rights, the parent-child relationship. Financial constraints and the "undercurrent of traditional values in opposition to children's rights" are cited, as well. The concept of children's rights has received little attention in the United States.

International Law

Basically, the Universal Declaration of Human Rights is seen as a basis for all international legal standards for children's rights today. There are several conventions and laws that address children's rights around the world. A number of current and historical documents affect those rights, including the 1923 Declaration of the Rights of the Child, drafted by Eglantyne Jebb and her sister Dorothy Buxton in London, England in 1919, endorsed by the League of Nations and adopted by the United Nations in 1946. It later served as the basis for the Convention on the Rights of the Child.

Convention on the Rights of the Child

The United Nations' 1989 Convention on the Rights of the Child, (CRC), is the first legally binding international instrument to incorporate the full range of human rights—civil, cultural, economic, political and social rights. Its implementation is monitored by the Committee on the Rights of the Child. National governments that ratify it commit themselves to protecting and ensuring children's rights, and agree to hold themselves accountable for this commitment before the international community. The CRC is the most widely ratified human rights treaty with 190 ratifications. Somalia and the USA are the only two countries which have not ratified the CRC. The CRC is based on four core principles, namely the principle of non discrimination, the best interests of the child, the right to life, survival and development, and considering the views of the child in decisions which affect them (according to their age and maturity).

The CRC, along with international criminal accountability mechanisms such as the International Criminal Court, the Yugoslavia and Rwanda Tribunals, and the Special Court for Sierra Leone, is said to have significantly increased the profile of children's rights worldwide.

Vienna Declaration and Programme of Action

Vienna Declaration and Programme of Action urges at Section II para 47, all nations to undertake measures to the maximum extent of their available resources, with the support of international cooperation, to achieve the goals in the World Summit Plan of Action. And calls on States to integrate the Convention on the Rights of the Child into their national action plans. By means of these national action plans and through international efforts, particular priority should be placed on reducing infant and maternal morality rates, reducing malnutrition and illiteracy rates and providing access to safe drinking water and basic education. Whenever so called for, national plans of action should be devised to combat devastating emergencies resulting from natural disasters and armed conflicts and the equally grave problem of children in extreme poverty.

Further para 48 urges all states, with the support of international cooperation, to address the acute problem of children under especially difficult circumstances. Exploitation and abuse of children should be actively combated, including by addressing their root causes. Effective measures are required against female infanticide, harmful child labour, sale of children and organs, child prostitution, child pornography, as well as other forms of sexual abuse. This gave an influence to adoptions of Optional Protocol on the Involvement of Children in Armed Conflict and Optional Protocol on the Sale of Children, Child Prostitution and Child Pornography.

Enforcement

A variety of enforcement organizations and mechanisms exist to ensure children's rights. They include the Child Rights Caucus for the United Nations General Assembly Special Session on Children. It was set up to promote full implementation and compliance with the Convention on the Rights of the Child, and to ensure that child rights were given priority during the UN General Assembly Special Session on Children and its Preparatory process. The United Nations Human Rights Council was created "with the hope that it could be more objective, credible and efficient in denouncing human rights violations worldwide than the highly politicized Commission on Human Rights."

The NGO Group for the Convention on the Rights of the Child is a coalition of international non-governmental organisations originally formed in 1983 to facilitate the implementation of the United Nations Convention on the Rights of the Child. Many countries around the world have children's rights ombudspeople or children's commissioners whose official, governmental duty is to represent the interests of the public by investigating and addressing complaints reported by individual citizens regarding children's rights. Children's ombudspeople can also work for a corporation, a newspaper, an NGO, or even for the general public.

United States Law

Children are generally afforded the basic rights embodied

by the Constitution, as enshrined by the Fourteenth Amendment to the United States Constitution. The Equal Protection Clause of that amendment is to apply to children, born within a marriage or not, but excludes children not yet born. This was reinforced by the landmark US Supreme Court decision of *In re Gault*. In this trial 15-year-old Gerald Gault of Arizona was taken into custody by local police after being accused of making an obscene telephone call. He was detained and committed to the Arizona State Industrial School until he reached the age of 21 for making an obscene phone call to an adult neighbor. In an 8–1 decision, the Court ruled that in hearings which could result in commitment to an institution, people under the age of 18 have the right to notice and counsel, to question witnesses, and to protection against self-incrimination. The Court found that the procedures used in Gault's hearing met none of these requirements.

There are other concerns in the United States regarding children's rights. The American Academy of Adoption Attorneys is concerned with children's rights to a safe, supportive and stable family structure. Their position on children's rights in adoption cases states that, "children have a constitutionally based liberty interest in the protection of their established families, rights which are at least equal to, and we believe outweigh, the rights of others who would claim a 'possessory' interest in these children." Other issues raised in American children's rights advocacy include children's rights to inheritance in same-sex marriages and particular rights for youth.

A report filed by the President of the INGO Conference of the Council of Europe, Annelise Oeschger finds that children and their parents are subject to United Nations, European Union and UNICEF human rights violations. Of particular concern is the German (and Austrian) agency, Jugendamt (German: Youth office) that often unfairly allows for unchecked government control of the parent-child relationship, which have resulted in harm including torture, degrading, cruel treatment and has led to childrens' death. The problem is complicated by the nearly "unlimited power" of the Jugendamt officers, with no processes to review or resolve inappropriate or harmful treatment. By German law, Jugendamt officers are protected against prosecution. Jugendamt (JA) officers span of control is seen in cases that go to family court where experts testimony may be overturned by lesser educated or experienced JA officers; In more than 90% of the cases the JA officer's recommendation is accepted by family court. Officers have also disregarded family court decisions, such as when to return children to their parents, without repercussions. Germany has not recognized related child-welfare decisions made by the European Parliamentary Court that have sought to protect or resolve children and parental rights violations.

CHILDREN'S RIGHTS MOVEMENT

The Children's Rights Movement is a historical and modern movement committed to the acknowledgment, expansion, and/or regression of the rights of children around the world. While the historical definition of child has varied, the United Nations Convention on the Rights of the Child explains, "A child is any human being below the age of eighteen years, unless under the law applicable to the child, majority is attained earlier." There are no definitions of other terms used to describe young people such as "adolescents", "teenagers" or "youth" in international law.

Thomas Spence's *The Rights of Infants* (1796) is an early English-language assertion of the natural rights of children.

In the USA, the Children's Rights Movement was born in the 19th century with the orphan train. In the big cities, when a child's parents died or were extremely poor, the child frequently had to go to work to support himself and/or his family. Boys generally became factory or coal workers, and girls became prostitutes or saloon girls, or else went to work in a sweat shop. All of these jobs paid only starvation wages.

In 1852, Massachusetts required children to attend school. In 1853, Charles Brace founded the *Children's Aid Society*, which worked hard to take street children in. The following year, the children were placed on a train headed for the West, where they were adopted, and often given work. By 1929, the orphan train stopped running altogether, but its principles lived on.

The *National Child Labor Committee*, an organization dedicated to the abolition of all child labor, was formed in the 1890s. It managed to pass one law, which was struck down by the Supreme Court two years later for violating a child's right to contract his work. In 1924, Congress attempted to pass a constitutional amendment that would authorize a national child labor law. This measure was blocked, and the bill was eventually dropped. It took the Great Depression to end child labor nationwide; adults had become so desperate for jobs that they would work for the same wage as children. In 1938, President Franklin D. Roosevelt signed the Fair Labor Standards Act which, amongst other things, placed limits on many forms of child labor.

Now that child labor had been effectively eradicated in parts of the world, the movement turned to other things, but it again stalled when World War II broke out and children and women began to enter the work force once more. With millions of adults at war, the children were needed to help keep the country running. In Europe, children served as couriers, intelligence collectors, and other underground resistance workers in opposition to Hitler's regime.

Present

In the early 20th century, moves began to promote the idea of children's rights as distinct from those of adults and as requiring explicit recognition. The Polish educationalist Janusz Korczak wrote of the rights of children in his book

How to Love a Child (Warsaw, 1919); a later book was entitled *The Child's Right to Respect* (Warsaw, 1929). In 1917, following the Russian Revolution, the Moscow branch of the organization Proletkult produced a Declaration of Children's Rights. However, the first effective attempt to promote children's rights was the Declaration of the Rights of the Child, drafted by Eglantyne Jebb in 1923 and adopted by the League of Nations in 1924. This was accepted by the United Nations on its formation and updated in 1959, and replaced with a more extensive UN Convention on the Rights of the Child in 1989.

From the formation of the United Nations in the 1940s and extending to present day, the Children's Rights Movement has become global in focus. While the situation of children in the United States has become grave, children around the world have increasingly become engaged in illegal, forced child labor, genital mutilation, military service, and sex trafficking. Several international organizations have rallied to the assistance of children. They include Save the Children, Free the Children, and the Children's Defense Fund.

The Child Rights Information Network, or CRIN, formed in 1983, is the group of 1,600 non-governmental organizations from around the world which advocate for the implementation of the Convention on the Rights of the Child. Organization's report on their countries' progress towards implementation, as do governments that have ratified the Convention. Every 5 years reporting to the United Nations Committee on the Rights of the Child is required for governments.

United States

While there is a long history of children's rights in the U.S., scholars contend that there is no "golden age". Many children's rights advocates in the U.S. today advocate for a smaller agenda than their international peers. Groups predominately focus on child abuse and neglect, child fatalities, foster care, youth aging out of foster care, preventing foster care placement, and adoption. A longstanding movement promoting youth rights in the United States has made substantial gains in the past.

United Kingdom

Anti-Children's Rights propagandists often raise the spectre of Rights without Responsibilities. The Children's Rights Movement assert that it is rather the case that children have rights which adults, states and government have a responsibility to uphold. The UK maintains a position that UNCRC is not legally enforceable and is hence 'aspirational' only - albeit a 2003 ECHR ruling states: "The human rights of children and the standards to which all governments must aspire in realizing these rights for all children are set out in the Convention on the Rights of the Child." (Extract from Sahin v Germany, Grand Chamber judgment of the ECHR, July 8, 2003). 18 years after ratification, the four Children's Commissioners in the devolved administrations have united in calling for adoption of the Convention into domestic legislation, making children's rights legally enforceable.

Several countries have created an institute of children's rights commissioner or ombudsman, the first being Norway in 1981. Others include Finland, Sweden and Ukraine, which was the first country worldwide to install a child in that post when Ivan Cherevko and Julia Kruk became the first children's rights ombudsmen in late 2005.

As already mentioned United Nations Convention on the Rights of the Child has outlined a standard premise for the children's rights movement and has been ratified by all but two nations - the United States and Somalia. Somalia's inability to sign the Convention is attributed to their lack of governmental structure. The US administration under Bush has opposed ratifying the Convention because of "serious political and legal concerns that it conflicts with U.S. policies on the central role of parents, sovereignty, and state and local law." However, the new administration may be taking another look at it. And also Optional Protocol on the

Involvement of Children in Armed Conflict against military use of children and Optional Protocol on the Sale of Children, Child Prostitution and Child Pornography against sale of children, child prostitution and child pornography.

REFERENCES

Ahearn, D., Holzer, B. with Andrews, L. (2000, 2007) *Children's Rights Law: A Career Guide*. Harvard Law School. Retrieved 2/23/08.

Arts, K, Popvoski, V, *et al.* (2006) *International Criminal Accountability and the Rights of Children*. "From Peace to Justice Series". London: Cambridge University Press.

Bandman, B. (1999) *Children's Right to Freedom, Care, and Enlightenment*. Routledge. p 67.

Brennan, S. and Noggle, R. (1997) "The Moral Status of Children: Children's Rights, Parent's Rights, and Family Justice," *Social Theory and Practice. 23.*

Brownlie, J. and Anderson, S. (2006) "'Beyond Anti-Smacking': Rethinking parent–child relations," *Childhood. 13*(4) p 479-498.

Calkins, C.F. (1972) "Reviewed Work: Children's Rights: Toward the Liberation of the Child by Paul Adams", *Peabody Journal of Education. 49*(4). p. 327.

Covell, K. and Howe, R.B. (2001) *The Challenge of Children's Rights for Canada. Wilfrid Laurier University Press.* p 158.

Cutting, E. (1999) "Giving Parents a Voice: A Children's Rights Issue," *Rightlines. 2* ERIC #ED428855.

DeLamater, J.D. (2003) *Handbook of Social Psychology*. Springer. p 150.

Franklin, B. (2001) *The new handbook of children's rights: comparative policy and practice*. Routledge. p 19.

Freeman, M. (2000) "The Future of Children's Rights," *Children & Society. 14*(4) p 277-93.

Jenks, C. (1996) "Conceptual limitations," *Childhood*. New York: Routledge. p 43.

Joseph M. Hawes, *The Children's Rights Movement: A History of Advoccacy and Protection* (Boston: Twayne Publishers, 1991).

Kaslow, FW (1990) Children who sue parents: A new form of family homicide? *Journal of Marital and Family Therapy. 16*(2) p 151–163.

Lansdown, G. "Children's welfare and children's rights," in Hendrick, H. (2005) *Child Welfare And Social Policy: An Essential Reader*. The Policy Press. p. 117

Lansdown, G. (1994). "Children's rights," in B. Mayall (ed.) *Children's childhood: Observed and experienced*. London: The Falmer Press. p 33.

Lansdown, G. (1994). "Children's rights," in B. Mayall (ed.) *Children's childhood: Observed and experienced.* London: The Falmer Press. p 34.

Lansdown, G. (1994). "Children's rights," in B. Mayall (ed.) *Children's childhood: Observed and experienced.* London: The Falmer Press. (p 33-34).

League for Children's Rights Individual UPR Submission: Germany. February 2009. Submitted by Bündnis RECHTE für KINDER e.V. and supported by President of the INGO Conference of the Council of Europe, Annelise Oeschger. Retrieved December 27, 2011.

Mangold, S.V. (2002) "Transgressing the Border Between Protection and Empowerment for Domestic Violence Victims and Older Children: Empowerment as Protection in the Foster Care System," New England School of Law. Retrieved 4/3/08.

Mason, M.A. (2005) "The U.S. and the international children's rights crusade: leader or laggard?" *Journal of Social History.* Summer.

Rodham, H. (1973). "Children Under the Law". Harvard Educational Review 43: 487–514.

Rojas Flores, J. (2007) "The rights of the child in Chile: an historical view, 1910-1930", *Historia. 40.*3. Special issue.

Starr, RH (1975) *Children's Rights: Countering the Opposition.* Paper presented at the 83rd Annual Meeting of the American Psychological Association in Chicago, Illinois, Aug. 30-Sept. 3, 1975. ERIC ID# ED121416.

Thorne, B. (1987) "Re-Visioning Women and Social Change: Where Are the Children?" *Gender & Society 1*(1) p. 85–109.

3

Towards Holistic Development of Children

> "There are really valid reasons why it is unwise to shield children from all violence, terror, sorrow, and death in their reading. The way to cure a child of fear of the dark is not to deny the existence of dark, but to walk with him in the dark and show him by example the restful quiet of it, and show him, too, how to avoid breaking his neck by stumbling over something he can't see."
>
> —*Roberg G. Mood*

It is pertinent to note that each individual child has a right to a happy, healthy and stimulating environment. It is parents and other caregivers at home who are primarily responsible for the growth and development of young children, and there is a growing realisation in our country that this important role, especially in disadvantaged communities, needs to be supported by integrated interventions for holistic child care. There is also a growing awareness that investment in young children contributes significantly in alleviating poverty in the developing world. In India, developmental programmes for infants under three years of age have not received as much attention as early childhood education for 3 to 5 year olds, which also needs to be expanded. The findings of medical and educational research - that the brain responds most to very early experience, that the environment influences brain function, and that half of all mental development potential is established before the age of four - these findings are yet to be translated into a more general awareness about the developmental significance of the first three years, and the crucial role of parents and other caregivers in promoting holistic development at this age. There is no doubt that in all families, even the most impoverished, babies and young children receive a great deal of love and concern.

Most parents understand the need for physical care for their children. A recent survey conducted by the Centre For Learning Resources revealed, however, that mothers and other family members in disadvantaged communities in our country generally believed that, learning begins only when a child starts going to school. They were unaware of the cognitive benefits of early stimulation,although some of their own caregiving practices promoted it. They also needed awareness of more ways of promoting such stimulation.

CLR's Supportive Interventions

Since its inception in 1984, the CLR has conducted inservice training programmes and developed a range of training and learning materials for early childhood education (ECE). It has assisted a number of organisations in improving their ECE programmes (Some examples : Deepalaya-New Delhi, Urban Basic Services- Andhra Pradesh, Bodh- Jaipur, AKES-Gujarat, U.P.

Basic Education Project, PLAN-Nepal, ICDS-Maharashtra, DPEP-Karnataka, Catholic Relief Services and several small voluntary organisations in Maharashtra). Building on this experience, the CLR is now also targetting the birth to 3 years age group. During the past few years, it has undertaken educational projects and research for promoting the holistic development of disadvantaged children in the birth to 3 years age group. It has given inputs to field-level NGOs who were providing services for women and young children in rural areas.

Assessing Needs Through Action-Research

Between 1997-99, a pilot project in collaboration with the Department of Women Child Development, Government of Maharashtra, and facilitated by UNICEF, was undertaken by the CLR in order to initiate interventions through ICDS. The action-research project, covering approximately 2000 families through 20 anganwadi centres was entitled " Mental Stimulation of Children in the Birth to 3 Years Age Group : A Home Based Approach Linked to Urban Anganwadi Centres of the ICDS". *

This project, alongwith its experience with some field-level NGOs, gave the CLR the opportunity to assess the existing situation of infant stimulation within poor urban and rural families and design a relevant training programme for field workers who could educate caregivers about psychosocial stimulation in infancy. Monitoring of caregivers' meetings and making home visits provided valuable feedback. The most significant learning was that messages related to psychosocial development had to be integrated with messages for child health and nutrition at the field level. There is a considerable amount of training materials in our country on child health and nutrition, and good sourcebooks on holistic child care.

However, the CLR realised that a need still existed for tangible communication (EC) materials with a focus on home-based care and psychosocial stimulation in infancy - materials which can be used not only for reference, but which can be interpreted easily and used directly by field workers or trainers in developmental programmes for women, children and their families.

Developing an Education Package for Caregivers

To address this need, the CLR developed in 1999-2000, an education package for parents and other caregivers, entitled "Right From The Start". The manual and accompanying visual materials comprising the package promote holistic home-based child care in disadvantaged communities. The interactive easy-to-use communication approach of the package helps field workers with modest literacy skills to convey basic messages related to health care, nutrition and emotional well-being, while emphasising the young child's need for psychosocial stimulation.

Special Features of the Education Package

"Right From The Start" endorses sound traditional caregiving practices. Taking into account the many constraints of time and space faced by mothers and other caregivers in poor families, the package suggests several realistic possibilities of when and how caregivers can interact with babies and toddlers and provide them with a stimulating environment, using the often scarce resources in their homes. The everyday spoken language of caregivers is used in all the communication materials, which are presently in Marathi. * In the entire communication approach to be followed by field workers, an illiterate and semi-literate audience has been kept in mind. Theoretical information is kept to the minimum necessary for them to understand the importance of certain adult-child interactions.

Components of the Education Package

Manual - This contains 18 modules, for conducting periodic meetings with caregivers of children in the birth to 3 years age group. Each module provides the necessary content and background information for a particular meeting, as well as step-by-step guidelines for conducting the meeting. In the entire communication approach, a semi-literate / illiterate audience has been kept in mind.

Discussion Photographs - Each set highlights the developmental characteristics of a particular age group of children, e.g. birth to 6 months, or 2 to 3 years. The modules explain how to use them during the meetings.

Other Visual Materials - Puzzles and Discussion Pictures. These help to make the meetings participatory and encourage caregivers to analyses issues for themselves.

Handouts for Caregivers - These are given to participants at the end of each meeting. They contain illustrated reminders of some of the main messages discussed at the meeting.

Focus Areas of the Package

"Right From The Start" incorporates the following key areas in early childhood care and development for the birth to 3 years age group:

- Understanding holistic development
- Role of the caregiver
- Developmental milestones and developmental characteristics of children at different stages
- Appropriate nutrition
- Prime messages in child health
- Promoting emotional well-being
- Significance of play and psychosocial stimulation
- Promoting caregiver-child interaction
- Age appropriate stimulation activities
- Play materials for young children
- Advocating gender equity
- Age-appropriate discipline and a safe environment

Regional Language Editions

Since the package is designed for direct use by field workers in rural areas and urban slums, regional language editions of the components will be required. The entire package is presently available in Marathi. The CLR is inviting inquires from non-governmental and government agencies interested in commissioning other regional language editions. A Hindi and Gujarati edition is already planned.

Training Programmes For Home-Based Child Care

The CLR has started training programmes for field workers and trainers involved with caregivers of children under 3 years of age. Through its training programmes and materials, the CLR is attempting to help families to stretch their understanding of what 'caregiving' means, and help them to develop intimate and responsive relationships with their babies and toddlers.

CLR Action Research Project

'Mental Stimulation of Children in the Birth - 3Years Age Group:

A Home-Based Approach Linked to Urban Anganwadi Centres of the ICDS'

Rationale

ICDS is one of the world's largest outreach programmes attempting to achieve major national health and nutrition goals for young children. Its preschool component also contributes to the national goal of universal primary education. It is an accepted feet, however, that psychosocial development and mental stimulation in the earliest years (birth -3) remain upto now largely neglected in this massive scheme. A consultation with the Department of Women and Child Development (DWCD), Government of Maharashtra, was fecilitated by UNICEF, in order to initiate interventions through ICDS for the psychosocial development of underprivileged children in this age group.

A pilot project was undertaken by the CLR in collaboration with the DWCD, entitled "Mental Stimulation of Children in the Birth-3 Years Age Group: A Home-Based Approach Linked to Urban Anganwadi Centres of the ICDS."

Objectives

The objectives of the project were to:

- *Assess* the situation of mental stimulation of children in the birth-3 age group.
- *Analyse* the information and design training inputs for caregivers and ICDS functionaries.
- *Train* ICDS functionaries to educate caregivers for early mental stimulation.

- *Monitor* the education sessions conducted by ICDS functionaries and the implementation in homes.
- *Provide feedback* for policy formulation.

Target Group

The pilot project was conducted in two ICDS projects in Pune city-Project 1 and Project 2 -which have 111 and 100 anganwadis respectively. These are divided into four supervisory sectors, each sector comprising of approximately 20-30 anganwadis under one supervisor. Two supervisory sectors, in Yerawada and Hadapsar respectively, were selected. Ten anganwadis were selected from each sector. Therefore, 20 anganwadi workers (AWWs) and two supervisors participated in the project. The selection of supervisory sectors and anganwadis was done by the DWCD, based on availability of space for meetings and ability and enthusiasm of the AWWs.

Time Frame

Project duration: fourteen months. Phase I lasted for 2 months and Phase II for 12 months.

Project Activities

Phase 1: Collection and Analysis of Data

The following three tasks were completed in the first phase:

1. House-to-house survey of the anganwadi areas to identify:
    - homes with children in the birth-3 years age group, as also pregnant women
    - actual caregivers of these children
2. Survey of selected homes to find out the existing mental stimulation activities that are naturally a part of daily routines in the homes.
3. Analysis of data to inform Phase II

Findings

The findings of both the house-to-house survey and observations in selected homes have been reported at length in the project report which is available with the CLR. The findings cover the following:

- Number of households and number of children in the birth - 3 years age group.
- Identification of primary caregivers and secondary caregivers; also working mothers
- Caregiver- child interaction during bathing the child
- Interaction during feeding
- Interaction while putting the child to bed
- Interaction while playing and other activities
- Attitude of caregivers towards psychosocial stimulation of the child at home

In what way did Phase I inform Phase II......

1. *It confirmed that caregivers require educational inputs for stimulation of infants in the home.*
2. *It helped to identify the extent and nature of traditional caregiving practices promoting stimulation.*
3. *It suggested several possibilities of interaction. It helped to identify the actual everyday objects available as possible play material in disadvantaged homes.*
4. *It pointed to four regular daily routines during which stimulating interaction could take place i.e. bathing, feeding, putting the child to bed, and play.*
5. *It pointed to the need to find strategies to target fathers as secondary caregivers.*
6. *It revealed that mothers are not the only*

caregivers. Therefore, the educational meetings should also target other female relatives in the household e.g. grandmothers and school-going siblings.

7. *It revealed that social tensions and family problems may be uppermost in the minds of many caregivers. That psychosocial stimulation is a crucial need for young children needed to be strongly and effectively conveyed.*

Phase II: Interventions

The following activities were carried out in Phase II:

1. Designing educational inputs for caregivers.
2. Designing and conducting the training programme for field-level 'communicators' (ICDS functionaries).
3. Monitoring the education sessions for caregivers conducted by ICDS functionaries
4. Monitoring home-visits by ICDS functionaries between education sessions, to motivate caregivers to participate and to observe implementation of messages received during the education sessions.

[*Courtesy:* CRL, Pune]

4

National Plan of Action on Children

> "Parents' discipline should be based on four Fs: firmness, fondness, frankness, and fairness. Parents who cannot say no to a child often rear offspring who have con-tempt for authority. Parents who get too angry over minor infractions often fail to instill a sense of discipline in their children."
>
> —*Alexander Martin*

Introduction

1. Right through the ages, care for children has been one of the causes to which Indian policy has remained committed. In independent India, this commitment was enshrined in our Constitutional provisions. The Constitution of India in its Directive Principles of State Policy pledges that "the State shall, in particular, direct its policy towards securing..... that the health and strength of workers, men and women, and the tender age of children, are not abused and that citizens are not forced by economic necessity to enter avocations unsuited to their age or strength; that children are given opportunities and facilities to develop in a healthy manner and in conditions of freedom and dignity and that childhood and youth are protected against exploitation and against moral and material abandonment." As a follow up to this commitment, Government of India adopted a National Policy for Children in 1974 which reaffirms the Constitutional provisions and declared that "it shall be

the policy of the State to provide adequate services to children, both before and after birth and through the period of growth, to ensure their full physical, mental and social development. The State shall progressively increase the scope of such services so that, within a reasonable time, all children in the country enjoy optimum conditions for their balanced growth." The specific measures to be adopted towards the attainment of these objectives are in Appendix-11.

2. Since independence human resource development programmes focussed on maternal and child health, nutrition and education have occupied an important place in India's efforts to raise the living standards of its poor. Successive Five Year Plans have provided the wherewithal to deal with these issues. In the last decade of this century, dramatic technological development particularly in health, nutrition and related spheres have opened up new vistas of opportunities for redeeming our age-old pledges to the cause of children. It is against this backdrop that India joined the comity of nations in the successive reaffirmations of global commitment to the cause of children in 1989-90. The UN Convention on the Rights of the Child in November 1989, the World Conference on Education for All at Jomtien in March 1990, the global consultation on Water and Sanitation in September, 1990, the World Summit on Children in the autumn of 1990 and the SAARC Summit on Children soon after the World Summit were all part of this reaffirmation process which transcended national barriers. India is a signatory to the World Declaration (September 1990) in the Survival, Protection and Development of Children and the Plan of Action for implementing it.

3. This National Plan of Action represents India's response to the unprecedented opportunities which the nineties decade has ushered in both through technology as well as a global ambience for redeeming India's age-old pledges. This Plan of Action identifies quantifiable targets in terms of major as well as supporting sectoral goals (detailed in Appendix-I of this Plan of Action) representing the needs and aspirations

of almost over 300 million children of India in the spheres of health, nutrition, education and related aspects of social support.

Situation of Children

4. Overall, basic indicators on children show a positive trend. For example, the infant mortality rate fell almost steadily from 146 per 1000 live births in 1960 to 80 per 1000 live births in 1990. The prevalence of severe and moderate degrees of malnutrition among children declined for the period 1974 to 1989. Primary school enrolment rates increased from 38 per cent in 1951 to 94 per cent in 1989. The availability of drinking water in rural areas improved significantly and by the end of the VII Plan (1989-90) 80 per cent of the population had been provided with potable water sources.

5. Major problems however, remain intertwined with the factors of poverty. For instance, about 30 per cent of the children born each year have low birth weight. The major causes of infant mortality viz. diarrhoea, pneumonia, neonatal tetanus and measles, are often aggravated by malnutrition. Protein-energy malnutrition affect about 52.5 per cent of all children below six years, with approximately a tenth of them suffering from severe malnutrition. Deficiency in Vitamin A causes, possibly among other adverse effects, an estimated 40,000 new cases of nutritional blindness each year. Some 70 per cent of women of child-bearing age suffer from chronic anaemia. Iodine deficiency is more widespread than hitherto suspected.

6. While there is a primary school in nearly all villages, problems remain in basic education, including lack of access to school, low achievement, high drop-out and repeater rates and severe disparities between socio-economic groups, geographic regions, rural and urban areas and between the boys and girls. While, Since 1951, the basic education system has grown into one of the largest in the world with an enrolment, in 1989, of about 131 million children reaching out to about 94 per cent of the population, the retention has

remained unsatisfactory with a drop out rate of 45 per cent for classes I-V and 60 per cent for classes I-VIII. At the end of the 1980's the overall picture shows that the gross enrolment ration for primary education has reached about 100 per cent however, the major reviews have advocated that targets for universalisation of elementary education should no longer be set in terms of additional enrolments for the country as a whole; disaggregated target setting policy has to be followed.

7. Inadequacy in the availability of safe drinking water, improper disposal of human excreta, solid and liquid water leading to unfavourable environmental conditions and lack of personal hygiene have been one of the major causes of disease and disability among children. In 13 States drinking water sources, are contaminated with excess fluoride causing thousands of children affected with dental and skeletal fluorosis. Excess nitrate in drinking water sources which might cause, Blue baby to children, is also observed in a number of States. About 15 lakh children die due to dehydration caused by Diarrhoea. As regards sanitation, the situation continues to be unsatisfactory as only about 3 per cent of the rural households are provided with low cost on-site human excreta disposal facilities as in 1990. Therefore, ensuring drinking water quality and strengthening its supportive linkages with health, and sanitation remain major challenges.

8. Children in especially difficult circumstances are without adequate family support and they may not be accessible through regular services. The rapid urbanisation, increased rural urban migration do have an adverse effect on children.

9. The situation of girls and boys differs significantly according to most indicators, particularly in the norther States.

10. The state of children hinges on the conditions of women. Most social indicators, including the sex ratio (929 females per 1000 males) and literacy (39 for females and 64

for males for all ages), point to the depressed situation of women. Socio-cultural bases, combined with poverty, weight heavily on women who marry early and bear children young, and who work excessively long hours in the home and outside, with unequal access to health and nutrition, educational and other opportunities, as well as insufficient legal protection and social and political participation.

11. The goals for children are promoted necessarly within the broader framework of national development planning. The key social development objectives of the 8th Five Year Plan (1992-97), are population control, employment generation and basic human needs particularly health care, literacy including elementary education and drinking water which is closely linked to sanitation.

12. Most of the recommendations of the World Summit for children contained in the Plan of Action for Survival, Protection and Development of Children were found relevant however, wherever these appeared to be too ambitious or beyond the resources likely to be available for these sectors, have been modified.

I. Health

A. Child Health

I.A.1. Major Goal: Reduction of infant mortality rate to less than 60 per thousand live births and reduction of child mortality rate to less than 10 by 2000 A.D.

I.A.1.1 Objectives

(i) Eradication of poliomyelitis by the year 2000;

(ii) Elimination of neonatal tetanus by 1995;

(iii) Reduction by 95 per cent in measles deaths and reduction by 90 per cent of measles cases compared to pre-immunisation levels by 1995;

(iv) Achievement and maintenance of high level of immunisation coverage at a level of 100 per cent of infants and against tetanus for women of child bearing age;

(v) Reduction by 50 per cent in deaths due to diarrhoea in children under the age of 5 years and 25 per cent of diarrhoea in incidence rate;

(vi) To endeavour to reduce mortality rates due to ARI among children under 5 by 40 per cent by 2000 A.D. from the present level.

I.A.1.2. Strategies

The basic instruments for achieving the stated objectives are.

I.A.1.2.1. Objective

Eradication of poliomyelitis by the year 2000 A.D.

Activities

(a) Strengthening of the existing primary health care infrastructure.

(b) Intensification of mortality/morbidity specific interventions by using the opportunity created by the Universal Immunisation Programme (UIP) in terms of continued contact with the mothers and young children for implementing a UIP Plus package of services combining immunisation with other basic MCH interventions, progressively with view to universalising this package by 2000 A.D. The components of this package would be:

For Young Children

Newborn Care, (below 3 years): Primary immunisation by 12 months. Administration of Vitamin 'A' Pneumonia therapy and control of diarrhoeal diseases.

For Mothers

Antenatal Care, Immunisation against tetanus. Anaemia control during pregnancy. Screening and referral of high risk cases, care at birth and promoting spacing and timing of births.

I.A.1.2.2. Objective

Elimination of neonatal tetanus by 1995.

Activities

The thrust would be control of vaccine preventable diseases and documentation of 'zero' levels of neonatal tetanus in a phased manner. The elimination of neonatal tetanus implies immunisation combined with better birth practices, with safe motherhood.

I.A.I.2.3. Objective

Reduction by 95 per cent in measles deaths and reduction by 90 per cent of measles cases compared to pre-immunisation levels by 1995.

Activities

(a) Specific measures will be taken to consolidate and maintain levels of immunisation coverage, and to step up immunisation where coverage is low.

(b) The immunisation services will be reprogrammed towards polio eradication, elimination of neo-natal tetanus and decrease in cases as well as deaths due to measles.

(c) Appropriate measures will be taken towards the provision of essential items in adequate quantities such as vaccines, syringes and needles and immunisation cards so that coverage level and quality of services do not drop.

(d) Field supervision would be improved.

I.A.1.2.4. Objective

To achieve and maintain Immunisation coverage at a level of 100 per cent of infants.

Activities

(a) "UIP Plus" package consisting of immunisation, control of diarrhoeal diseases, pneumonia diagnosis and therapy. Vitamin 'A' and Newborn care will be implemented.

(b) Initiatives will be taken to ensure that essential supplies and drugs are made available regularly and on time at health units.

(c) Steps will be taken to create demand for the package of services for children.

(d) Inter-sectoral programmes such as integrated Child Development Services (ICDS), Urban Basic Services (UBS) and Development of Women and Children in Rural Areas.(DWCRA) will be activated for reaching young children and women with set of complementary interventions for improved mother and child survival and health.

(e) Emphasis will be given to decentralised planning and implementation using the district as a Unit.

(f) A holistic approach aimed at better health of women and children through affordable means, by adopting 'high risk' criteria will be the strategy.

I.A.1.2.5. Objective

Fifty per cent reduction in deaths due to diarrhoea in children under the age of 5 years and 25 per cent reduction in the diarrhoea incidence rate.

Activities

(a) The training communication, supply and logistic support for village level care for diarrhoeal diseases will be made operational university in districts in a phased manner.

(b) Case management will be improved at hospital referral centres, more specially, district hospitals and primary health centres.

(c) Diarrhoea Training Units at Medical College Hospitals will be set up.

(d) Intensive efforts will be made to network with

non-governmental organisations and educate the people.

(e) Several village level outlets will be used for provision of ORS and a communication campaign will be launched to create a demand for ORS.

(f) Special emphasis will be given to home management of diarrhoea-mothers and families will be trained by both interpersonal as well as mass media to adopt appropriate practices for correct case management at home and timely referral.

(g) The wide-spread infrastructure of other programmes like rural development, urban development and ICDS, will be used to extend the outreach for education of mothers on appropriate home care.

LA.1.2.6. Objective

To endeavour to reduce mortality rates due to ARI among children under-5 by 40 per cent by 2000 A.D. from the present level.

Activities

(a) The rational use of antibiotics for treatment would be ensured in a phased manner in all the districts.

(b) Training of the PHC doctors and female Para-medical Health Workers.

(c) Communication support to the Programme from within and outside the health sector to enhance 'use of services.

(d) Action for promoting home management of mild infection and timely referral to a health worker or an appropriate facility.

(e) Proper care management by health workers involving correct assessment and treatment

B. Maternal Health

I.B.1. Major Goal

Between 1990 & the year 2000, reduction of maternal mortality rate by half.

I.B.1.1. Objectives for Women's Health and Education.

(i) Special attention to health and nutrition of the female child and to pregnant and lactating women;

(ii) Access by all couples to information and services to prevent pregnancies that are too early, too closely spaced, too late or too many;

(iii) Access by all pregnant women to prenatal care, trained attendants during child birth and referral facilities for high risk pregnancies and obstetric emergencies;

(iv) Universal access to primary education with special emphasis for girls and accelerated literacy programmes for women.

Strategies for Objectives (i) to (iii)

(i) A minimum of three ante-natal check-ups during pregnancy will be arranged to detect high risk groups and for complete immunisation.

(ii) Pregnant women will be immunised against tetanus.

(iii) There will be anaemia control activities during pregnancy.

(iv) Spacing and timing of births will be promoted.

(v) Specific measures, necessary for consolidation and maintenance of high levels of immunisation coverage and stepping up the coverage wherever they are identified to be low, will be taken up.

(vi) Health care units will be strengthened as first level referral to deal with obstetric emergencies.

(vii) Traditional birth attendants would be trained/ retrained to ensure that deliveries are conducted by IBAs.

(viii) Measures will be taken for skill development of medical and paramedical workers for institutionalising the referral systems for management of obstetric complications.

(ix) The work routines of the key worker i.e. the Auxiliary Nurse and Midwife (ANM) would be rationalised be re-defining her tasks so as to include only those activities that will result in preventing deaths.

(x) There would be increasing involvement of Voluntary Workers at village, level, so that ANM is able to concentrate on delivery of services, such as immunisation, Vitamin A Prophylaxis, iron and Folk Acid and on the management of cases of diarrhoea, anaemia, pneumonia and pregnancy.

(xi) District will be the Unit for planning and management of services. District plans would be developed based on the infrastructure and facilities for providing services. It will be the unit for supply, social mobilisation and monitoring.

(xii) Steps will be taken to create demand for the package of services for mothers and children.

(xiii) Girls between the age of 13 and 20 shall be reached through an exposure programme which covers the basic of safe motherhood and timing of births.

(xiv) Tetanus toxide will be promoted during the pre-pregnancy period for adolescents and this event will be used to identify anaemia among adolescents and to appraise their health status.

(xv) Activities to promote health growth of adolescent girls will be implemented, such as supplementary iron.

I.B. 1.2. Objective

Universal access to Primary Education with special emphasis for girls and accelerated literacy programme for women.

Activities

(a) Provision of additional infrastructure will be made.

(b) Efforts will be made to enhance the self image and self-confidence of women and to enable them to recognise their contribution to the economy as producers and workers.

(c) Provide women with the necessary support structures and an informal learning environment to create time for education.

(d) To create an environment which young girls and adolescent girls working in their homes in agriculture and elsewhere get an opportunity for formal education.

(e) Provision of primary education facilities in unserved or undeserved areas as alternative models for education such as use of village women, low cost hostel facilities, widening primary education opportunities for girls, increasing the number of women teachers, providing residential accommodation for women teachers.

(f) Create Child Care Service to release girls from sibling care responsibilities for participating in educational activities.

(g) Improve access to drinking water and sanitation facilities within the community as a whole so as to lessen the burden of household chores among girls and women.

(h) Create non-formal educational facilities as an alternative to formal schools.

(i) Community mobilisation and involvement efforts.

(j) Provision of suitable incentives and facilities for girls.

Nutrition

II. 1. Major Goal

Between 1990 and the year 2000 A.D. reduction in severe and moderate malnutrition among under-5 children by half.

Objectives

(i) Reduction in severe as well as moderate malnutrition among under-5 children by half of 1990 levels,

(ii) Reduction in incidence of low birth weight (2.5 kg. or less) babies,

(iii) Reduction of iron deficiency anaemia in women,

(iv) Control of iodine deficiency disorders,

(v) Control of Vitamin A deficiency and its consequences including blindness,

(vi) Empowerment of all women to breast-feed their children exclusively for four to six months and to continue breast-feeding with complementary food, well into the second year,

(vii) Growth promotion and its regular monitoring to be institutionalised by the end of the 1990s,

(viii) Dissemination of knowledge and supporting services to increase food production to ensure household food security.

II. 1.2.1. Objective

Reduction in severe, as well as moderate malnutrition among under-5 children by half of 1990 levels..

Activities

(a) Expansion of nutritional intervention net through ICDS so as to cover all vulnerable children in the age group of 0-6 years.

(b) A concerted effort to bring about appropriate behavioural changes among the mothers through existing programmes such as the Integrated Child Development Services (ICDS), safe motherhood, Urban Basic Services (UBS), Development of Women and Children in Rural Areas (DWCRA) and programmes of Food and Nutrition Board.

(c) Improvement in growth monitoring between the age group 0-3 years with closer involvement of mothers to be taken.

(d) Encouraging small family norms and adequate spacing through intensive family welfare and motivational measures so that availability of food is sufficient

(e) Involving the community in the identification of problems and management of nutrition programmes and related interventions such as health education, involvement of women in food production and processing activities and other employment generation activities.

(f) Emphasis on women's employment and education particularly nutrition and health education.

(g) Convergence of services by strengthening linkages between the concerned sectors like agriculture, food, health, women and child development, education, rural development, urban development, etc.

(h) Creation of conducive environment by providing safe drinking water, clean environment, immunisation service, health care etc.

II. 1.2.2. Reduction in Incidence of Low Birth Weight (2.5 kg or less) Babies.

(a) Nutritional communication will be developed to create greater awareness of nutritional problems and their solutions amongst the general public.

(b) Nutrition education, will be closely linked to activities like immunisation, Oral Rehydration Therapy, promotion of breast-feeding, birth spacing and female education.

(c) Appropriate low cost nutrition foods for Supplementary Feeding Programme at ICDS Project areas will be developed.

(d) Correct dietary habits for improving nutritional levels through behavioural change will be promoted.

(e) Ante-natal and post-natal care to women for preventing low birth weight babies will be provided.

(f) Ensuring better nutritional coverage of expectant mothers right from the 1st trimester for the major period of lactation.

II. 1.23. Reduction of Iron Deficiency (Anaemia) in Women

(a) Consumption of iron-rich foods will be improved and promoted through Integrated Child Development Services (ICDS), Maternal and Child Health (MCH), Urban Basic Services (UBS) & Development of Women and Children in Rural Areas (DWCRA) Programmes.

(b) "UIP Plus" package will include control of iron deficiency anaemia through ensuring iron supplements to pregnant women.

(c) All pregnant women and 50 per cent of young anaemic children will be covered with iron and folic acid.

(d) Improved quality, packaging and distribution of iron and folic acid tablets will be ensured.

II. 1.2.4. Control of iodine deficiency disorder

(a) A comprehensive strategy for control of IDD will be developed. The distribution and consumption of iodised salt will be promoted through various channels.

(b) Research and development studies will be sponsored to assess the feasibility of double fortification of salt with iron and iodine.

(c) The strategy for communication and training as well as management information system for MCH services and ICDS will include component of control of iodine deficiency disorders.

(d) Linkages with contact points such as the immunisation programme and ICDS will be developed to ensure maximum coverage of control of iodine deficiency.

II. 1.2.5. Control of Vitamin A deficiency and its consequences including blindness.

(a) National Prophylaxis programmes for control of Vitamin-A deficiency will be effectively implemented.

(b) Emphasis will be placed on efforts to improve diets rich in Vitamin-A.

(c) Capacity to produce more Vitamin-A concentrates will be expanded further with a view of effectively implement national policy on Vitamin-A programmes.

(d) Administration of Vitamin-A to all children between nine months to three years, as well as treatment doses to infants and young children following each attack of diarrhoea, measles and acute respiratory infection will be ensured.

(e) Utilisation of existing nutrition, health services will be further improved.

(f) Breast-feeding and better infant feeding practices will be promoted.

II. 1.2.6. Empowerment of all women to breastfeed their children exclusively for four to six months and to continue breast-feeding with complementary food, well into the second year.

Activities

(a) Awareness amongst functionaries of various programmes such as Integrated Child Development Services (ICDS), Urban Basic Services (UBS), Development of Women and Children in Rural Areas (DWCRA) workers, health workers, doctors, families and mothers will be created for promotion of breastfeeding.

(b) Health workers will be trained to promote appropriate lactation management and breastfeeding.

(c) The communication strategy for child survival and health will include breastfeeding as an integral component.

II. 1.2.7. Growth promotion and its regular monitoring to be institutionalised by the end of the 1990s.

Activities

(a) Growth promotion and its regular monitoring will be an integral component of health and nutrition programme. Involvement of women in growth monitoring would be encouraged.

(b) Gross root level workers of Integrated Child Development Services (ICDS), Urban Basic Services (UBS) & Maternal and Child Health (MCH) will be trained in growth monitoring.

II. 1.2.8. Dissemination of knowledge and supporting services to increase food, production to ensure household food security.

Activities

(a) Ensured nutritional adequacy by increasing the production and increasing availability of nutritionally rich food.

(b) Utilisation of the available food resources through the application of effective food technology.

(c) Food availability to ail-specially weaker sections to ensure household food security.

(d) Strengthening of Public Distribution System and Incorporation of nutritional consideration in Public Distribution System and in all poverty alleviation programmes.

III. Water and Sanitation

III. 1. Major Goal

Universal access to safe drinking water and improved access to sanitary means of excreta disposal.

III. 1.1. The norms for providing drinking water in rural areas and those pertaining to sanitation are as follows:

Water

(i) To arrange potable water source within a distance of one Km. horizontal and 50 metres elevation difference.

(ii) One spot source or standpost for 100/150 persons.

Sanitation

A package linked with the demand and need and with a differential beneficiary contribution.

III. 1.2. Objectives

(i) To provide the entire rural population with potable water supplies @.40 Ipcd.

(ii) To cover 10 per cent of population with sanitary facilities by the year 1997.

(iii) To eradicate guineaworm disease by 1995.

(iv) To provide safe water with fluoride content within tolerable limits by the year 2000 A.D.

Activities

Water

(i) Coverage of remaining no-source villages.

(ii) Coverage of habitation with special emphasis on Scheduled Castes (SC) Scheduled Tribes (ST).

(iii) Augmentation of service level,

(iv) Improved operation and maintenance,

(v) Quality improvement.

(vi) Water resource management and prevention of failure of sources,

(vii) Health Education, Community Participation and Awareness programme.

Sanitation

(i) Taking sanitation as a package.

(ii) Adoption of a demand and need based approach to make the programme a people's movement.

(iii) Establishment/Strengthening of State/District Sanitation Cell.

(iv) Intensive district programming.

(v) Development of appropriate delivery system.

(vi) Adoption of an appropriate IEC strategy,

(vii) .Empowerment of women on improved sanitary practices.

(viii) Co-ordination with other related programmes.

(ix) Involvement of community, voluntary organisations and NGO's.

(x) R & D to develop appropriate area specific low cost technology to suit different geo-hydrological conditions.

IV. Education

1. Major Goal

Universal enrolment, retention, minimum level of learning, reduction of disparities and universalisation of effective access of schooling.

1.1 Objectives

(a) Universal enrolment of all children including girls, using both full time formal schools & part time non-formal arrangements.

(b) Reduction of drop-out rate between class I to V and I to VIII from the existing 45 per cent and 60 per cent to 20 per cent and 40 per cent respectively.

(c) Achievement of minimum level of learning by approximately all children at the primary level and introduction of this concept at the upper primary stage on large scale.

(d) Reduction in disparities by emphasis on girl's education and special measures for children belonging to Scheduled Castes/Scheduled Tribes (SC/ST).

(e) Expansion of Early Childhood Development activities

including appropriate low cost family and community based interventions.

(f) Universalisation of effective access to schooling. The two broad strategy frames for reaching the targets are:

   (i) Micro-planning through people's participation and decentralisation; and

   (ii) Introducing Minimum Levels of Learning (MLL) in schools. Micro-planning would be area specific, population specific through mobilising community participation, decentralising educational administration, reorienting and strengthening local administrative and resource support systems, ascertaining educational requirements of the area, bringing to school all children who can be enrolled, seeing that all children regularly and actually participate in primary education and planning for the improvement of the schools. Minimum levels of learning strategy focuses on learning acquisition in schools & outcomes and is proposed to be implemented through an assessment of the existing level of learning achievement; a definition of the MLL for the area and the time frame for achieving it; reorientation of teaching practices to competency based teaching; introduction of evaluation of student learning; review of textbooks and re-revision, if required and the provision of inputs to improve learning acquisition to the MLL including provision of physical facilities, teacher training, supervision, evaluation etc.

Objectives

(i) Universal enrolment of all children including girls, using both full-time formal schools and part-time non-formal arrangements.

(ii) Reduction of drop-out rate between classes I to V and I to VIII from the existing 45 per cent and 60 per cent to 20 per cent and 40 per cent respectively.

(iii) Achievement of minimum level of learning by approximately all children at the primary level and introduction of this concept at the upper primary stage on a large scale.

Activities

(i) Seek people's participation by arousing them to the need for education, creating a sense of responsibility towards children's education and empowering them to take accountability of the school system.

(ii) Community mobilisation and environment building will use folk and traditional media, people-to-people discussions, group meetings etc.

(iii) Village Education Committees (VEC) will be formed in each village, with adequate representation of women and the disadvantaged groups, to interact with teachers, provide support to the school, ensure regular participation of all children through parental motivation and take accountability over the running of the school.

(iv) Efforts will be made to involve suitable NGOs or field level organisations/project management and to strengthen them adequately for such provision of support to the VEC. Capacity building at local levels will be an essential activity in all areas adopting the microplanning strategy.

(v) In the next 5 years, efforts will be made to shift educational planning to the district level, by setting up District Boards of Education and District Institutes of Education and Training (DIETs).

(vi) An effective MIS will be created to provide information required for decision making, planning and management to the district levels and information for monitoring district plans of the State and national levels.

(vii) The accent during the next 5 years will be upon opening of new primary schools according to the norm in the unserved habitation, and the provision of non-formal education programmes for smaller habitations and for children who cannot benefit from the school system.

(viii) In order to increase enrolment at the upper primary stage, the infrastructure at the upper primary stage will be considerably expanded.

(ix) The existing norms of providing an upper primary school within 3 km walking distance, for all practical purposes, denies education at this level to a large section of girls. This norm will be relaxed and the existing ratio between primary and upper primary schools will be improved to 1:2.

(x) Provisions will be made to add additional rooms and provide additional teachers to existing schools through the scheme of Operation Blackboard and the existing norm for the minimum-2 classrooms and 2 teachers-will be raised successively.

(xi) Non-formal education (NFE) will be developed as a complementary system to the formal education system, such that there will be parallel administrative apparatus for NFE; greater attention to the interests of the learner, and its quality will be comparable with formal system and conform to the MLLs laid down for the primary stage.

(xii) Mechanisms will be developed for entry from non-formal to the formal system and vice-versa.

(xiii) The NFE programme will include a substantial enhancement in costs per learner, on the reasoning that it is not possible to reach out to the marginalised groups with marginalised funding.

(xiv) In order to ensure that regular in-service training and academic supervision is provided to teachers, a beginning will be made to create Teachers' Centres for every cluster of 20-25 primary schools to provide a forum to teachers to get together, discuss issues, resolve common problems and receive academic guidance to build upon their capacities and skills.

(xv) A programme of in-service training of teachers will be coordinated by this block level team to infuse a level of competency in teachers to effectively transact curriculum and textual material.

(xvi) At the State level the existing Councils of Educational Research and Training (SCERT) will be restructured so that they become institutions principally for resource support to elementary education, non-formal education and early childhood education.

(xvii) Resource support through researches and studies will be encouraged by continuing support to Colleges of Teacher Education and Institutes of Advanced Studies in Education.

(xviii) District specific plans will be developed as projects, with specific activities, clearly defined responsibilities, definite time-schedules and district specific targets. Apart from effective UEE the goals of each project will include the reduction of existing disparities in educational access, the provision of alternative systems of comparable standards to the disadvantaged groups, bringing about a substantial improvement in the quality of schooling facilities, obtaining a genuine community involvement in running of schools, and building up local level capacity to ensure effective decentralisation of educational planning.

(xix) A monitoring system will be developed to obtain timely and reliable information on enrolment, retention, completion and achievement. In addition the qualitative monitoring of achievement will also be introduced.

Objective

For Reduction in Disparities: by Emphasis on girls' education, and Special measures for children belonging to Scheduled Castes/Scheduled Tribes (SC/ST).

Activities

(i) An effort will be made to provide a comprehensive system of incentives and support services for girls and children of the economically weaker sections of society.

(ii) A provision of services such as establishment of Day Care Centres for pre-school children and infants, provision of free uniforms, text books and stationary as attendance incentives to girls, coordination of support services such as drinking water, fodder and fuel.

(iii) A similar incentive as appropriate, will be provided in the non-formal system, also, in addition to provision of free text books and stationery to all NFE learners.

Objective

For Early Childhood Education:

Expansion of early childhood development activities including appropriate low-cost family and community based interventions.

Activities

(i) During the VIII Five Year Plan, the ICDS will be expanded and will be the main scheme for early childhood activities.

(ii) The contribution of ICDS to basic education will be enhanced by adjusting it to school timings and by strengthening its educational component.

(iii) Other existing schemes for early childhood education will be improved so as to become holistic schemes providing nutrition and health care as well as education.

(iv) Emphasis and efforts will be made on training of personnel, production of sufficient play way learning material for children, creation of a system of monitoring and supervision; and development of effective rapport between ECCE and community and the primary school.

2. Major Goal

Achievement of adult literacy rate of 80 per cent (which is of mastery level) in the age group 15-35, with emphasis on female literacy.

Objective

Imparting of functional literacy so as to enable the learners to achieve self-reliance in literacy and numeracy, to make them aware of the causes of their deprivation and move towards amelioration of their condition through organisation and participation in the process of development, acquire skills to improve their economic status and general well-being and imbibe in themselves the values of national integration, conservation of the environment, women's equality, observance of small family, norms etc.

Activities

(i) Continuous environment building conducive to literacy through jathas, rallies, media and communication, traditional folk art forms etc. and securing a national consensus among all sections of the society including political parties, representatives of the people, etc.

(ii) Organising mass campaigns for total literacy which are area-specific, time-bound, volunteer-based, cost-effective and result-oriented in 345 districts of the country up to 1997.

(iii) Involving a large number of voluntary agencies in

programmes relating to eradication of illiteracy in a campaign mode.

(iv) Increasing involvement of students in schools and colleges, nonstudent unemployed youth, youth clubs, teachers and representatives of other non-governmental organisations, etc. in literacy activities.

(v) Use of the new teaching/learning pedagogy evolved under the technique of Improved Pace and Content of Learning (IPCL) providing for reduced duration of learning, heightened motivation of learners and an in-built mechanism for self evaluation by the learners.

(vi) Strengthening of the academic and technical resource support to the programme by establishment of additional State Resource Centres, Regional Resource Centres, District Resource Units, etc.

(vii) Making adequate arrangements for training of all field functionaries and timely supply of teaching/learning material.

(viii) Developing, designing and patenting various techno-pedagogic imputs for improving the pace and quality of the programme by pursuing the findings of the scientific and technological research.

(ix) Imparting literacy to the learners in their spoken language and supply of district-specific and culture-specific teaching-learning material to heighten and sustain the motivation of the learners.

(x) Increasing emphasis on coverage of women learners by setting up of large number of adult education centres exclusively for women, involvement of women volunteers in substantial numbers, giving preference to women in the matter of appointment of Preraks of Jana Shikshan Nilayams and providing teaching-learning material relevant to the needs of women.

(xi) Launching of post-literacy campaigns immediately

alter the conclusion of the total literacy campaigns so as to prevent the risk of the neoliterates relapsing into illiteracy.

(xii) Establishment of Jana Shikshan Nilayams for providing facilities of post literacy and continuing education in areas which are covered by the programmes other than total literacy campaigns.

(xiii) Establishing linkages with other developmental programmes like health & family welfare, conservation of environment, agriculture, rural development, etc.

(xiv) Evolving suitable mechanisms for continuous monitoring and evaluation of the programme at various levels.

(xv) Continuous research on Adult Education Programmes of all descriptions to make the Programmes more meaningful and effective.

V. Children in Especially Difficult Circumstances

V. 1. Major Goal

Improved protection of children in especially difficult circumstances.

Objective

Provision of protection of children in especially difficult circumstances and tackle the root cause leading to such situations. This would include children in following categories; physically handicapped; mentally handicapped; drug addicts, victims of natural and man-made disasters, refugee children, street children, slum & migrant children, orphans & destitutes; children suffering from AIDS, children of parents with AIDS & AIDS orphans, children of prostitutes & child prostitutes & juvenile delinquents and child labour.

Activities

(a) Efforts will be made to build linkages between existing Child Care Institutions and communities and society at large for greater community participation for children in especially difficult circumstances.

(b) The staff of institutions and implementing bodies will receive pre and in-service training.

(c) Interventions for street children and other children in difficult circumstances will be made with active support of the Municipal Bodies, Health Department, Education Department etc. through relevant programmes in the social service sector.

(d) To strengthen prevention of Child Labour, emphasis will be on compulsory education for all children and on strengthening anti-poverty and development programmes and focussing them on at risk families.

(e) Voluntary Organisations will be assisted to undertake projects for welfare and development of children in especially difficult circumstances.

(f) National Child Labour Policy 1987 will be taken up more vigorously for implementation.

V. 2. Goal

To assist children affected by one or more disabilities, having no access to proper rehabilitative services and especially to lift up the status of those most marginalised.

Strategy

To set in motion advocacy and services for disabled children in two main phases;

(i) to help people understand that disabilities are a commonly occurring phenomenon with specific causes and that their own endeavour can help reduce both the incidence of disability and its severity, and

(ii) to ensure that people recognise that rehabilitation is possible and to help inform them about how to receive those services.

Activities

There will be four main priority areas for work in child disability fields:

(i) To enable and to mobilise all central areas in social services beginning at the community level to assist in the process of primary detection, prevention, and rehabilitation;

(ii) To create a cadre of very well trained workers ranging in expertise from early detection and rehabilitation at the community level, to those who have the most up-to-date skills; to make sure that the best quality care is made available at the most appropriate level for disabled people.

(iii) Communication efforts needed at all levels will be made to alert people to the fact the disabilities need not become handicaps, that they can be prevented, that when they occur they need not unnecessarily prevent an individual from being a full member of the community. Good communication campaigns which can also deeply change people's attitude towards disablement and towards people who are affected by a disability or handicap will be launched. Communications campaigns and strategies that can ensure that all of those who can affect and improve the quality of services available to the disabled people see this as an appropriate priority and take suitable action.

(iv) To support and encourage the various institutions, both government and nongovernmental which are putting new ideas into practice in the field of disability. Support would be given to these institutions to help them implement them and to measure the impact and effect of these strategies.

V. 3.

Amongst children who are neglected, abandoned or

rendered destitute a large number of children are handicapped and special provisions need to be made for such children. The aims and objectives for the handicapped children would, therefore, be:

(i) to provide residential facilities to children suffering from severe/profound mental retardation or cerebral palsy;

(ii) to provice facilities for care and protection, maintenance, treatment and rehabilitation for such, children;

(iii) to provide pre-vocational training in accordance with the needs and capabilities of the clildren; and

(iv) to workout development based approaches, strategies, programmes and services for the children.

VI. Girl Child

VI. 1. Objective

To remove the gender bias and to improve the status of the girl child in society so as to provide her with equal opportunities for her survival and development to her full potential.

Activities

Keeping in view the prevailing gender bias and disadvantaged status of the female child, special programmes and interventions directed at the girl child are required to ensure her an equal status and a better quality of life. These should take into account her peculiar social, cultural and economic mileau which would require special-strategies and interventions. The Action Plan for the SAARC Decade ill of the Girl Child includes programmes of both advocacy and action in the developmental areas of health, education, welfare etc. A special focus will be ** child survival. The Action Plan is proposed to be implemented in close collaboration with the concerned Central Ministries, State Governments and Non-governmental Organisations.

VII. Adolescent Girls

VII. l.Objective

Programme for the adolescent girls should embrace a whole range of activities, such as nutrition, health, education, health and nutrition education, recreation, upgradation of home-based skills and promotion of her decision making capability as adolescent girls who are one of the most critical human resource, particularly at the village level with the following objectives:

(i) to cover girls in the age group of 11 to 18 years;

(ii) to improve the nutritional and health status of girls in this age groups;

(iii) to provide them the required literacy and numeracy skills, through the non-formal stream of education, to stipulate a desire for more social exposure and knowledge and to help them improve their decision making capabilities, especially on issues regarding their future.

(iv) to train and equip the girls to improve and upgrade home based skills;

(v) to promote awareness of health, hygiene, nutrition and family welfare, home management and child care, to take all other measures as would facilitate their marrying only after attaining the age of 18 and, if possible, even later.

VIII. Children and the Environment

VIII. 1. Objective

To conserve and protect environment so that it is conducive to the health and well being of the children.

Activities

Awareness should be created among children regarding important of protection of the environment and they should be encouraged to participate in activities aimed at prosection of environment.

IX. Women

Development of Women is intertwined with that of Children and constitutes a vital component, of a country's human resource development Therefore, these two target groups deserve special treatment in addition to their legitimate share from all other general developmental programmes. The major objective of all the developmental programmes, both women-specific and women-related has, therefore, been to ensure a state of 'well-being' and children particularly those of the weaker sections of society. In this direction, a special emphasis will be on developing Women's potential and main-streaming them into the national development process through enhanced access to skill development and income generation opportunities which are expected to provide not only the desired economic independence but also to help improve their quality of life as well as that of their families and children. The ongoing 27-beneficiary oriented schemes for women in the major developmental areas of health, education, family welfare, employment, rural and urban development, science and technology, etc. will be further strengthened by adding additionalities like safe motherhood, child survival, employment and income generation, etc.

X. Advocacy and People's Participation

Advocacy for the child as everyone's concern and advocacy with policy makers, planners, programme implementors at national and international levels for focus on the child will be integral to the achievement of the goals. In order to implement the plan of children, voluntary organisations, local bodies, religious institutions and political parties would be mobilised for achieving the above mentioned goals. All channels of communication, formal and informal including mass media would be utilised for wider dissemination of the urgent need for achieving goals concerning children.

XI. Resources

Efforts would be made to mobilise adequate financial resources for implementation of the National Plan of Action on Children.

XII. Monitoring and Evaluation

To monitor the quality of service and the attainment of goals, the country will select appropriate indicators for each of the stated goals, to be monitored by the relevant sectors at regular intervals. For this the following activities will be pursued:

(i) Development and monitoring of critical indicators for measuring progress towards the goals;
(ii) For participatory monitoring of programme coverage and results at the community level; and
(iii) Action research and field studies relevant to the achievement of the goals for children.

APPENDIX I

I. Health

A. Child Health I.

A. 1. Major Goal

Reduction of infant mortality rate to less than 50 per thousand live births and reduction of child mortality rate to less than 10 by 2000 A.D.

I.A. 1.1. Objectives

(i) Eradication of poliomyelitis by the year 2000;
(ii) Elemination of neonatal tatanus by 1995;
(iii) Reduction by 95 per cent in measles deaths and reduction by 90 per cent of measles cases compared to pre-immunisation levels by 1995;
(iv) Achievement and maintenance of high level of immunisation coverage at a level of 100 per cent of infants and against tetanus for women of child bearing age;
(v) Reduction by 50 per cent in deaths due to diarrhoea in children under the age of 5 years and 25 per cent of diarrhoea in incidence rate;

(vi) To endeavour to reduce mortality rates due to ARI among children under 5 by 40 per cent by 2000 A.D. from the present level.

B. Maternal Health

I.B. 1. Major Goal

Between 1990 and the year 2000 reduction of maternal mortality rateby half.

I.B. 1.1. Objectives for Women's Health and Education.

(i) Special attention to the health and nutrition of the female child and to preganant and women.

(ii) Access by all couples to information and services to prevent pregnancies that are too easy too closely spaced, too late or too many,

(iii) Access by all pregnant women to prenatal care, trained attendants during child birth and preferral facilities for high risk pregnancies and obstetric emergencies,

(iv) Universal access to primary education with special emphasis for girls and accelerated literacy programme for women.

Nutrition

II. 1. Major Goal

Betweeen 1990 and the year 2000 A.D. reduction in severe and moderate malnutrition among under-5 children by half.

Objectives

(i) Reduction in severe as well as moderate malnutrition among under-5 children by half of 1990 levels;

(ii) Reduction in incidence of low birth weight (2.5 kg of less) babies;

(iii) Reduction of iron deficiency anaemia in women;

(iv) Control of Iodine deficiency disorders;

(v) Control of Vitamin A deficiency and its consequences, including blindness;

(vi) Empowerment of all women to breastfeed their children exclusively for four to six months and to continue breast-feeding with complementary food, well into the second year;

(vii) Growth promotion and its regular monitoring to be institutionalised by the end of the 1990s;

(viii) Dissemination of knowledge and supporting services to increase food production to ensure household food security.

III. Water and Sanitation

III. 1. Major Goal

Universal access to safe drinking water and improved access to sanitary means of excreta disposal.

III. 1.1. The norms for providing drinking water in rural areas and those pertaining to sanitation are as follows:

Water

(i) To arrange potable water source within a distance of one Km. horizontal and 50 metres elevation difference.

(ii) One spot source or standpost for 100/150 persons.

Sanitation

A package linked with the demand and need and with a differential beneficiary contribution.

III. 1.2. Objectives

(i) To provide the entire rural population with potable water supplies @ 40 Ipcd.

(ii) To cover 10 per cent of population with sanitary facilities by the year 1997.

(iii) To eradicate guineaworm disease by 1995.

(iv) To provide safe water with fluoride content within tolerable limits by the year 2000 A.D.

IV. Education

IV. 1. Major Goal

Universal enrolment, retention, minimum level of learning, reduction of disparities and universalisation of effective access of schooling.

IV. 1.1. Objectives

(i) Universal enrolment of all children including girls, using both full time formal schools and part time non-formal arrangements.

(ii) Reduction of drop-out rate between class I to V and I to VIII from the existing 45 per cent and 60 per cent to 20 per cent and 40 per cent respectively.

(iii) Achievement of minimum level of learning by approximately all children at the primary level and introduction of this concept at the upper primary stage on large scale.

(iv) Reduction in disparities by emphasis on girl's education and special measures for children belonging to Scheduled Castes/Scheduled Tribes (SC/ST).

(v) Expansion of Early Childhood Development activities including appropriate low cost family and community based interventions.

(vi) Universalisation of effective access to schooling.

IV. 2. Major Goal

Achievement of adult literacy rate of 80 per cent (which is of mastery level) in the age-group 15-35, with emphasis on female literacy.

Objective

Imparting of functional literacy so as to enable the learners to achieve self-reliance in literacy and numeracy, to make them aware of the causes of their deprivation and move towards amelioration of their condition through organisation and participation in the process of development, acquire skills to improve their economic status and general well-being and imbibe in themselves the values of national 'integration, conservation of the environment, women's equality, observance of small family norms, etc.

V. Children In Especially Difficult Circumstances

V. 1. Major Goal

Improved protection of children in especially difficult circumstances.

Objective

Provision of protection of children in especially difficult circumstances and tackle the root cause leading to such situations. This would include children in following categories; physically handicapped; mentally handicapped; drug addicts, victims of natural and man-made disasters, refugee children, street children, slum and migrant children, orphans and destitutes; children suffering from AIDS, children of parents with AIDS and AIDS orphans, children of prostitutes and child prostitutes and juvenile delinquents and child labour.

V. 2. Goal

To assist children affected by one or more disabilities, having no access to proper rehabilitative services and especially to lift up the status of those most marginalised.

VI. Girl Child

VI. 1. Objective

To remove the gender bias and to improve the status of the girl child in society so as to provide her with equal opportunities for her survival and development to her full potential.

VII. Adolescent Girls

VII. 1. Objective

Programme for the adolescent girls should embrace a whole range of activities, such as nutrition, health, education, health and nutrition education, recreation, upgradation of home-based skills and promotion of her decision making capability as adolescent girls who are one of the most critical human resource, particularly at the village level as a part of the ICDS programme with the following objectives:

(i) to cover girls in the age group of 11 to 18 years;

(ii) to improve the nutritional and health status of gils in this age group;

(iii) to provide them the required literacy and numeracy skills, through the non-formal stream of education, to stipulate a desire for more social exposure and knowledge and to help them improve their decision making capabilities, especially on issues regarding their future;

(iv) to train and equip the girls to improve and upgrade home based skills;

(v) to promote awareness of health, hygiene, nutrition and family welfare, home management and child care, to take all other measures as would facilitate their marrying only after attaining the age of 18 and, if possible, even later.

VIII. Children and the Environment

1. Objective

To conserve and protect environment so that it is conducive to the health and well being of the children.

IX. Women

Existing programmes for women to raise their status in society with special emphasis on creation of employment opportunities for them would be strengthened keeping in view the direct impact of such programmes on the welfare of children.

X. Advocacy and People's Participation

Advocacy for the child as everyone's concern and advocacy with policy makers, planners, programme implementors at national and international levels for focus on the child will be integral to the achievement of the goals.

5

Issues in Child Protection Work

> "One reason why the upbringing of a child is so difficult nowadays is that no sooner have you taught it to talk than you have to start teaching it to keep quiet."

It is pertinent to note that in this chapter attention will be focused on issues currently being debated by researchers and practitioners in the field of child protection. Most will have been touched upon so far in the course of this book, but it is intended to develop them further here. Inevitably the issues considered are a selection from the wide range that exist. Choice has been determined as much as possible by relevance to mainstream statutory practice in the child protection field.

Merits and Demerits of Current Child Abuse Theory and Research

A considerable amount of theory and research has been discussed so far in this book. There are several problems associated with these contributions that should be considered by those who make use of them. First, as has been seen, child abuse is a topic that is dealt with by a wide range of professional and academic specialities. Child abuse research reflects the theory bases and ways of thinking of these different specialities. All have insights to offer, but there is need for careful weighing and assessment of the way in which these different interests and ideological standpoints influence the outcomes. A good example of this need is in the field of prediction studies considered in the previous chapter. Here, psychologists and medical doctors have adopted a view, with some empirical backing, that there is a measurable correlation between certain social and psychological factors and the propensity for abuse.

Their background and training, based largely on positivist research, problem-solving and notions of medical prevention and cure, lead them to use these criteria to select out potential abusers and focus treatment on them. From a sociological perspective, which sees child abuse as a socially constructed problem in which professionals are not disinterested observers, this type of approach is more open to question. The factors used by psychologists and others to identify potential abuse are by no means water-tight predictors and it is seen as more logical and moral to address those factors (for example, social stress) which are associated with abuse potential directly, rather than to use them to identify the potential abuser. Reference has been made throughout to the cultural context in which knowledge is derived and, therefore, to the relevance of that knowledge to different cultures.

The work of Korbin (1981) is particularly instructive on this issue. However, she considers sharply contrasting societies where the lesson seems more obvious. What about the transferability of such knowledge between more similar types of society? How useful, for instance, is knowledge about intervening into child sexual abuse allegations among middle-class families in Santa Clara County, California, to working with deprived and disadvantaged families in industrial inner-city areas in Britain (and vice versa)?

The answer is not as negative as it might seem at first sight, because such communities share some common features, not least the view that sexual abuse of children by adults is unacceptable, harmful and unlawful. However, there may be stark differences in the acceptability and viability of methods of intervention derived from these different settings.

Reports from other industrialized European countries point to a much less open approach to the issue of child sexual abuse within the family than in either the USA or Britain, reflecting the different cultures, mores and beliefs that prevail there. There has been a good deal of scepticism about the worth of much child abuse research as a result of issues such as those just considered. O'Hagan (1989), arguing from a relevance-to-practice point of view, and in the wake of the Cleveland crisis, where, as we have seen, social workers' adherence to new ideas on child sexual abuse was considered to be misguided, writes:

> All literature on child sexual abuse has contributed in varying degrees to our understanding. In this practitioner's view, however, very little of the literature is either relevant or helpful in equipping front-line professional staff to cope with the multiple problems created by the abuse. (O'Hagan 1989: 21)

O'Hagan proceeds to argue that this state of affairs is not the fault of the researchers, but of the subject matter. His view is that child sexual abuse is so multivarious in the forms that it takes that it is not easily categorized or harnessed by research-based knowledge. This seems to be somewhat pessimistic and at the same time suggests that the only way forward is to rely on practical wisdom and experience. My view is that although available research on child sexual abuse has considerable limitations, it is more useful than O'Hagan suggests. However, it has to be taken from a range of sources and carefully weighed and assessed for relevance to different contexts and different situations. There are dangers associated with unthinkingly following a single model derived from another context. The problem with such an approach seems to be whether social workers and other front-line professionals have the time, commitment and organizational support to develop the type of expertise that the more careful analytical approach being suggested here requires.

The child abuse knowledge base has been attacked from another quarter, that of researchers. Dingwall is very critical of the existing knowledge base:

> The amount of scientifically-validated research on child abuse and neglect is vanishingly small. The value of any sell-styled predictive check-list is negligible. Indeed such tools probably do more harm than good because of the way they further undervalue and undermine professional judgement. Child abuse research will not progress until the scientific and the social problems are clearly distinguished. The former are the topic for the mainstream of investigations into child development, family interaction and aggression within a variety of social sciences from sociology and anthropology through psychology and ethology. The latter are perhaps more distinctively the concern of the sociologist, the political scientist and the social historian. (Dingwall 1989: 51)

While this view has much to recommend it, the implication is that there is no place for research based more specifically on child abuse, because it is a socially defined phenomenon and, therefore, does not lend itself to scientific methods of investigation. This seems a rather harsh judgement. An alternative approach is to improve definitions and to be somewhat more modest about the validity and applicability of findings.

Parton argues for child abuse as a topic to be more firmly situated in the general child care field:

> We could gain considerably by recognising in research, policy and practice that child abuse is centrally concerned with debates about the way we bring up children and is hence centrally concerned about child care. In the process the message seems to be that child abuse and child protection practice

> should be located in child care practice. (Parton 1989: 74)

Again, this view has much merit. In the past, separating child abuse from other child care concerns has had the negative effect of it being seen as a distinct and pathological phenomenon far removed from what happens between parents and children in so-called normal families. The linkages between child abuse and general child care far outweigh the differences and certainly there are vast overlaps between what is defined as moderate abuse and what is seen as a family experiencing child care problems and difficulties.

Nevertheless, the focus on child abuse has had some positive outcomes. It has greatly increased our understanding of and sensitivity towards violence in the family and its potentially debilitating effect on future mental health. It is also argued (Belsky and Vondra 1989) that a focus on child abuse can provide insights into normative family behaviour. Finally, it is an area of work that has attracted greater political and financial attention than has generally been afforded children and family issues, albeit at a price.

Thus, while it is important to locate child abuse research in the wider field of child care, nevertheless there are, in my opinion, benefits to be gained from maintaining a separate focus. The knowledge base for child protection work has many deficiencies. Great care needs to be taken in applying such knowledge to practice, particularly with regard to approaches that make grandiose claims about predictability and effectiveness. Nevertheless, the insights and aids to thinking about child abuse provided by research are of considerable importance for practitioners trying to get to grips with the massive complexities of child protection work.

Both research and practice until recently have treated different forms of abuse as separate phenomena. This is particularly true in the case of physical and sexual abuse. Physical abuse has traditionally been seen both by practitioners and researchers as a parent, more usually mother, focused issue. The abused child has been seen as the object of abuse and has not been considered to be in need of particular attention, except in more extreme cases. In cases of sexual abuse, on the other hand, the child has been seen as the key person whose needs are paramount. More recently, partly as a result of increased concerns about emotional abuse, there has been a growing realization that all forms of abuse have similar psychological consequences for the child and more attention is now being paid to these and other factors that are common across the board.

This is a welcome trend, in that insights gained and lessons learned from the understanding of and intervention into one form of abuse can be transferred to and tested with others. The following two issues demonstrate the potential value of thinking laterally about physical and sexual abuse. First, it is generally accepted that every effort should be made to ensure that a child who has been sexually abused should not be doubly abused by being taken into care (unless necessary for her or his protection) and thus deprived of her or his normal support systems.

The same consideration should also be given to children who have been physically abused. The situation is more complex in such cases, as there may be greater uncertainty as to who the abuser is and it is statistically more probable than in the case of sexual abuse to be the mother (an additional problem is that she is also more likely to be the main carer). Nevertheless, where possible, consideration should be given under Schedule 2 (5) of the 1989 Children Act to encouraging alleged physical abusers to live elsewhere pending the outcome of an investigation.

Second, children who are physically abased should be listened to and their accounts given the same sort of credibility as those of children who have been sexually abused. This point is vividly brought homo in the Stephanie Fox inquiry report:

> In June, Stephanie is reported to have told both a Social Services receptionist and day nursery staff that a bruise on her forehead was done by her daddy. She was quite clear about what happened. Perhaps more thought might have been given to talking with the child. If this had been a suspected sexual abuse case, a 'disclosure interview' would have been a priority. Should we not also listen to physically abused children? (Fox 1990: para. 5.10, 30-1)

Focusing on similarities between forms of abuse in this way should influence practitioners to question their assumptions and working practices and to make the best use of their knowledge and experience. It is to be hoped that the more sensitive approach to child sexual abuse will have an influence on the style of intervention into physical abuse cases and also that the sharper child protection focus that now characterizes physical abuse work can, where appropriate, be applied in cases of sexual abuse. None of this should be construed as thinking that there are no differences between types of abuse at all. However, at this stage, it may be more constructive to develop practice around the commonalities.

There are signs that this is beginning to happen in British child protection practice. For instance, the 1991 *Working Together* guidelines (DoH 1991b) do not distinguish between physical and sexual abuse at all, though it is not made clear whether this is a deliberate attempt to view child abuse more holistically or not. Whatever the case, the message seems to be getting across that it is important to respond to the emotional needs of all abused children and all members of a family in which any form of abuse occurs, in contrast to the earlier dichotomous approach to physical and sexual abuse.

Contribution of the Feminist Perspective to Child Protection Policy and Practice

In fact, there can be no doubt that over the past ten

years the feminist perspective has made a major contribution to explaining why child abuse happens. It has provided an additional dimension to the long-running debate about whether psychological or sociological factors, such as emotional deprivation or class and poverty, lie at the root of child abuse and has reframed some of the important issues. Initially, the aim of the feminist perspective was to reinforce the message that, in the case of sexual abuse, men were, almost exclusively, the perpetrators and needed to accept responsibility for their actions. Sexual abuse was seen from this perspective as an abuse of male power over largely female children, a state of affairs brought about and maintained by patriarchal power relations at an institutional level.

This power imbalance was seen not only as an explanation of why sexual abuse happened, but also as a major factor in the way in which such abuse was responded to, and as responsible for the consequences that followed from it. Examples of these effects were the fact that abused children were often blamed for the abuse, as were mothers, and the strong resistance to the opening up of the problem from men who found the subject matter threatening to their own values. Generally, this line of argument has been negatively perceived and often seen as irrelevant to direct practice. However, it has so far served its purpose of raising awareness and exposing the issues. There are now other important messages emanating from the feminist perspective with more direct implications for practice as well as understanding. One of these is the view that forms of gender socialization play a significant part in the aetiology of child abuse and that there is, therefore, a pressing need to address this issue in child protection policy and practice.

As far as sexual abuse of children is concerned, it is argued that the way in which males are socialized both generally, i.e. with the belief that strength lies in coping with one's own emotions, and particularly sexually (see Coveney *et al.* 1984), i.e. with the expectation that they have stronger sexual needs than women and must be assertive and take control, creates the climate in which such abuse can happen. The picture seems to be somewhat different with regard to physical abuse because of the fact that men and women are, according to known statistics, equally likely to abuse children. The feminist perspective does not deny that women are violent to children, but emphasizes how gender socialization and beliefs about motherhood tend to implicate them more than is their due. More importantly, these factors help to create conditions that give male abusers the opportunity to evade responsibility for their actions.

There are clear implications of this line of thought for both child protection practitioners and child abuse researchers. First and foremost, the term 'parent' should not be synonymous with 'mother'. There is a need to separate out the interests of parents. The term 'parents' suggests shared values, concerns and interests. This may simply not be the case and is an assumption that needs testing out.

Second, social workers and researchers must make greater efforts to include men in their practice and studies. In both sexual abuse and physical abuse work, men are generally not the focus of attention. Male sexual abusers are coming into the frame more than before. However, there still seems to be little focus on men who are physically violent towards children. If, as the feminist perspective argues, the roots of violence to both children and women lie in male socialization, then the logical target for change is being systematically missed.

Reference has been repeatedly made throughout the previous chapters to the way in which research projects, particularly in the field of physical abuse, can also be gender-blind in their approach. In much research, as in practice, the focus has been on female parents because they are more visible and available in that they are more likely to attend family centres and clinics with their children and, in many cases, are the sole adult providing constant care. Third, there is a need for male and female child protection practitioners to examine their own socialization experiences and the impact that they have on the way in which they construe and respond to child abuse work. In Cleveland, for instance, it has been convincingly argued that a clash in perspective between a male-dominated police force and two determined, strong women, Marietta Higgs and Sue Richardson, was a major factor in the prolongation of the crisis there (Campbell 1988).

According to this account, these two women felt justified in taking a lead role in a problematic area in which they considered themselves to have particular expertise. The police, threatened by the issue of child sexual abuse anyway, found it hard to accept a role that was subordinate to women and, therefore, pathologized their approach. There is a need for much greater recognition of the fact that female health and welfare practitioners are far more likely as a result of their own socialization experiences to have greater awareness than their male counterparts of the way in which women are subjected to sexual oppression. This in turn is likely to result in greater potential for empathy and, therefore, makes women more likely as a whole to be more effective practitioners in such cases. This is generally recognized in social services departments and other involved agencies.

However, it is important that this does not result in sexual abuse work becoming woman's work alone. If this happens, then agencies are mirroring what happens in the wider society and reinforcing beliefs and behaviours that seem to lie at the heart of the problem. There is a need for more official recognition of this line of thought. For instance, the Cleveland report itself makes no mention of the part played by gender socialization in the conflict between professionals there. The overview of recent inquiry reports (DoH 1991b) comments on the lack of reference to gender issues in any of the reports between 1980 and 1989. The feminist perspective should be seen as having an important positive part to play in child protection work. It highlights major deficiencies in current practice and thinking that have persisted because the male perspective has been dominant.

Close examination of all mainstream areas of work from the feminist perspective offers new and positive possibilities for intervention.

Rights of Parents and the Needs of Children

Achieving what is considered to be a reasonable balance between parents' rights and the need to provide protection for children has, as was seen in Chapter 3, been a perennial problem for the state throughout the history of child abuse prevention. The state is highly ambivalent about its role with regard to child-parent relationships (see Donzelot 1980) and this ambivalence is highlighted when things go wrong within families.

Essentially, parents are considered to have overarching responsibilities for the care and upbringing of their children, and within certain parameters formally required by the state, such as the need to ensure that children attend school, they also have considerable powers. Children have few independent rights until they reach the age of 16. The prevailing view is that parents have the right to make important decisions on behalf of their children and that this right is only forfeitable if they fail to carry out their duties towards them in an acceptable way. Even then, the state is wary of intervention because of its commitment to the family as a form of socialization. Indeed, its child care policy is geared to and dependent on the maintenance of the two-parent family.

This is reflected in the fact that alternatives to the family, such as foster care and residential services for children, are generally poorly resourced and stigmatized. On the one hand, therefore, the state is keen to ensure that parents should not be discouraged from carrying out their socializing role by being too heavily penalized when things go wrong.

On the other hand, it cannot tolerate serious maltreatment of children. Thus agencies such as social services departments and the NSPCC are empowered to remove the rights of parents, but only as a last resort. It is of vital importance that health and welfare practitioners have an awareness of the complex nature of the politics of intervention into family life. Whereas in theory the powers that agencies such as social services departments and the NSPCC have invested in them are unequivocal and no direct mention is made of paying due regard to the rights of parents in the process, in practice, for the reasons outlined above, the rights of parents have always been major considerations in decisions made 'in the best interests of the child'.

The expectation placed on social workers and others is that they will get the balance right. Their task is to calculate the risks involved for children in each situation referred to them and balance this against what seems fair to parents. They must do this in the light of legal rules, community values and resource factors. Judging from recent history, social workers have been seen, at least in the public eye, to have failed to maintain the right sort of balance. In physical abuse cases, such as that of Jasmine Beckford, they have been seen to lean too heavily towards consideration of parents' rights. In sexual abuse cases such as those investigated in Cleveland and the Orkneys, social workers have been seen to be acting overzealously on behalf of children. The 1989 Children Act is the latest attempt to tackle the problems of getting the balance right. On the one hand, it has widened powers to take control of a child about whom there is serious concern and whose parents are being uncooperative. It has increased the range of cases in which a guardian *ad litem* will be appointed to act specifically on behalf of the child. It has also empowered state agencies to proceed against parents where they suspect that a child will be subjected to significant harm in the future, whereas the previous legislation required abuse to be happening in the present and strictly limited intervention into potential abuse to certain specific situations.'

On the other hand, the rights of parents to challenge such decisions have been expanded. They have been given full party rights in care proceedings and the right to appeal against emergency protection orders. Agencies have been placed under considerable pressure to clarify the reasons for their actions. Other shifts towards giving greater weight to parents' rights are to be found in the recent *Working Together* guidelines (DoH 1991b), particularly with respect to parental participation at case conferences. The new Children Act has been seen by some as a children's charter and by others as a bill of rights for parents. In fact it is a bit of both and this is indicative of a continuation of the perennial ambivalence of the state about intervention into parent-child relationships.

The 1989 Act provides no radical rethink of the problem of child protection. If anything, it makes the balancing act between preserving the rights of parents and protecting children rather more complex and difficult to achieve than before. Health and welfare practitioners must, therefore, continue to operate in a Janus-like way with one eye on the parents and one on the child. This is a reality that must be lived with pending wider social changes, such as the allocation of more rights of independence to children. Society should also acknowledge this task and the difficulty of getting it right.

Anti-racist Approach to Child Protection Work

Remarkably, a major issue that has not to date received much attention but is likely to loom large in the future is that of how the child protection system can respond adequately to the needs of children at risk in black families. There is very little information about how child protection workers are currently responding in these situations. Some early accounts of child care intervention into black families were paternalistic and judgemental in tone and suggested that black children were more likely to be deemed at risk than white children and therefore more likely to be taken into care (Fitzherbert 1967). More recent accounts have pointed to greater tentativeness about intervening in black families where there are concerns about child care (Stubbs 1989). This is borne out to some extent in recent inquiry reports (Beckford, Henry) which have intimated that social workers have overcompensated for the structural disadvantages experienced by black families in our society to the detriment of the needs of children within them. The vital issues seem to be the need to respond sensitively and in an antiracist way to black parents without overlooking the fact that their children might be at risk and in need of protection.

As far as possible, the approach required seems to be very similar to that needed for good work with white families, with the additional awareness of the racism that many black people experience both as a result of general structural disadvantage (see Brown 1984) and in their contact with agencies such as social services departments, most of which are largely controlled by white managers and geared to the needs of a white clientele (Rooney 1987). White workers can make mistakes (a) as *a* result of their ignorance of cultural differences, i.e. having no awareness of the impact of culture on child-rearing practices, and (b) as *a* result of making assumptions based on cultural stereotypes. White professionals working with black families where abuse is suspected should be able to consult with advisers from the same cultural background as the client and should have access to interpreters where necessary.

Attempts should be made to have at least one black professional at a case conference involving a black family. Chairpersons of conferences need particular awareness of an anti-racist perspective and must *be* able to check racist assumptions in this setting. Following the adoption of measures such as these to guard against racial bias, it is important to be clear in one's mind that steps may well have to be taken to remove black children from their parents for their protection. Anti-racism *is* about providing services that are positively sensitive to the needs of black people, including their children, and *as* with white families these needs may well be in conflict. This type of thinking must be applied to all aspects of child protection work, not just initial investigations. Thus, therapeutic intervention and the provision of facilities such as day nurseries and family centres need to share an anti-racist perspective and take as much action as possible to offset alienation of black families.

Child protection work provides a particular challenge in this aspect because it is an activity that is perceived as questioning the quality of care being provided for children. It therefore generates a good deal of hostility and conflict that needs to be carefully worked through. Race and culture issues complicate matters even further. The adoption of comprehensive and constructive anti-racist policies by agencies can provide a background from which this difficult area of work can be conducted positively and with greater confidence.

Inter-Professional Cooperation as a Strategy for Dealing with Child Abuse

All the public inquiry reports published since 1974 have pointed to inter-professional communication and cooperation as vital ingredients in successful child protection work, and in all the cases inquired into major deficiences have been found in this respect. The Maria Colwell report pinpointed this issue. The child protection system that was established following this inquiry was based on procedures aimed at maximizing cooperation between agencies, and this inter-disciplinary philosophy has persisted up until now.

Despite the formalization of such procedures, however, problems have continued to exist:

> A lesson to be learned from the 1980s is that inter-agency working is not easy, and not self evidently useful. This is not to say that an individual agency can go it alone, but separate viewpoints and confusion of roles, as well as the availability of multiple pathways for communication, are a recipe for muddle. (DoH 1991b: 41)

Stevenson (1989) analyses some of the reasons for this state of affairs and points to three possible reasons. First, there are problems with the structural organization of child protection work. Area Child Protection Committees (ACPCs), which are appointed to coordinate the input of agencies, are poorly resourced, have limited powers with regard to influencing the parent bodies of their constituent members and have an unrealistically wide range of functions.

A major problem lies in the fact that individual agencies can still carry out their statutory functions regardless of recommendations from ACPCs. Thus, in Cleveland, the various agencies interpreted their roles and duties for themselves and went their own ways regardless of ACPC guidelines. The Kimberley Carlile report defined the child protection system as 'a puppet with no-one to pull the strings' (p. 142). Second, different professional groupings find it very difficult to work cooperatively towards a common goal. There are a variety of reasons for this, such as the lack of inter-professional contact during training, the fact that different professions have different functions, goals and philosophies, different statuses and pay, and different levels of responsibility for and commitment to the child protection system. The third impediment to better inter-professional cooperation is the fact that there can be major differences in the views that individuals have about what should or should not be done about different children at risk regardless of their professional roles and responsibilities.

Individuals bring their own experiences of family life to case conferences, their own views of what constitutes adequate care of children and their own experience of other cases where intervention of a certain kind has either proved successful or not. Factors such as these can exacerbate the difficulties of cooperation already considered. None of these problems is insurmountable and there are signs of improved working together in many areas. Much depends on individuals developing trust and respecting each other's role requirements by working together on a series of cases and by taking part in shared training. Unfortunately this is not always possible. Continuity can be hampered by high staff turnover.

If staff do not specialize, they may come into contact with each other only rarely and on-the-job training is both expensive and time-consuming. The important question to be asked is whether the type of inter-professional coordination that we currently rely on is the best way to respond to child abuse, given the difficulties outlined above. Alternative approaches have received some consideration. The Kimberley Carlile report considered that there was a pressing need for change and outlined a range of proposals, including the setting up of permanent inter-disciplinary teams. It concluded that a thorough review of organizational structures was required:

> Whatever the way forward. . . the issues should not be avoided on the grounds of their scale and complexity if the safety of children demands that they should be addressed. . . . We should not demand that individuals from the relevant professions continue striving to improve their performance, and continue bearing the responsibility for their work unless we are satisfied they are working within an overall framework that maximises their skills. (Carlile 1987:143)

Further, a second more modest proposal came from the Cleveland report, namely that *ad hoc* inter-disciplinary Special Assessment Teams should be set up to assess problematic child sexual abuse allegations. These should consist of panels of medical, police and social services personnel with sufficient authority to institute preliminary investigations and to decide whether there is enough reason to proceed with a full investigation. Very few authorities have formally adopted this approach. However, as pointed out in Chapter 8, the notion of holding preliminary meetings or strategy discussions seems to be a widespread practice and is recommended in the recent *Working Together* guidelines (DoH 1991b). Much time and effort is spent on getting professionals to cooperate over child protection investigations. Despite some of the advances and improvements referred to above, problems and difficulties continue to exist. There is a clear need for more research into the feasibility of different types of inter-professional cooperation in the field of child protection.

Future of The Child Protection System

It is pertinent to note that child protection work has developed over the past 25 years from being a relatively minor offshoot of general child care work to being a major aspect of the work of social services departments and of several large voluntary child care agencies. It takes up increasingly more time and resources of the health, education, police and probation services.

Officially recognized abuse as measured by child abuse registration quadrupled in the second half of the 1980s. Much of this expansion was in areas of abuse that had either been newly discovered, such as sexual and emotional abuse, or redefined as abusive, such as neglect. In addition, there was a massive increase in the numbers of children on registers because of serious, but undefined, concerns on the part of involved professionals. By 1991, over 40,000 children were on registers, with approximately 20,000 being registered each year and 20,000 de-registered. There are few signs of a slowdown in the number of cases being referred. New concerns are constantly being brought to the fore, such as abuse of children in boarding schools and residential homes, abuse of disabled children etc. This broadening of focus represents a major shift from the original concern for babies and young children being physically injured by their parents. All agencies have invested considerable resources in child protection work. Social services departments have a variety of organizational responses. Some employ specialist teams. Some retain general child care teams in name, but in practice child protection work takes up the bulk of time. Many intake teams in social services departments spend much of their time dealing with child protection investigations. Nearly all social services departments employ child protection coordinators to chair conferences, manage the administration of the system, including registers, and service Area Child Protection Committees.

Considerable secretarial support is required for administering the system and maintaining communication links within and between agencies. Police departments are devoting more time and resources to child protection work, including specialist teams in some areas, interviewing suites and joint training initiatives. Some education departments employ child protection advisers. Every school is expected to have a teacher with designated responsibility for dealing with child protection issues. Many health departments employ nursing managers with a remit to oversee all child protection work. Many major hospitals provide specialist child protection services. This represents a major shift from the period in the early 1970s when the only agency with a specific remit and structure for dealing with what was then termed child neglect was the NSPCC. How effective this response has been is a question that has not been subjected to any systematic research. Any attempt to assess the current situation has, therefore, to be an impressionistic one.

Most cases referred into the child protection system are now subjected to much closer initial scrutiny than those referred in the previous era, when there was greater reliance on the judgement of individual workers and their own initiative in working with other agencies. Most assessments of situations and decisions about intervention, planning and the allocation of resources are now subject to inter-disciplinary airing and backed by the authority of middle-level managers. The focus is much more on systematic sifting of cases to decide on risk and to ensure that the risk is managed by an accountable person, usually a social worker.

In addition there is much more attention being paid to a more open approach to parents and to involving them more in the decision-making process. The benefits of this approach are that plans are shared and agreed on by all agency workers, and in some cases parents, responsibility is clearly allocated and at a later date cases are reviewed. Cooperation between agencies is built in right at the start of the intervention so that in theory the sorts of difficulties highlighted by inquiries in the past regarding poor inter-agency communication should not exist.

The effect produced is more one of a team approach that can enhance confidence. One weakness lies in the fact that decisions are often made by those who are not responsible for resource allocation. Thus plans may be agreed, but the resources required to implement them, such as personnel time, a day centre place or suitable accommodation for a family, may not be available. Second, there is little room for discretion or flexibility in such a system, either with regard to what cases should be put into it or with regard to changing plans without case conference approval (see Corby 1987:135; Finkelhor and Zellman 1991).

This can lead to rigid implementation of plans, with the emphasis on authority rather than on responsiveness to the client and the changing situation. An overriding problem is whether the system that now exists can cope with the demands placed upon it. As was seen in Chapter 4, the American system is seen by some to be so overloaded that while cases are all formally processed and plans are formulated, in many cases they are not put into action (Krugman 1991). Thus the system gives the impression of dealing with cases, but the actual input is limited.

These and other problems seem to exist in the British system as well. Pressure of cases can result in delayed conferences, low levels of attendance and long gaps between conferences and the sending out of minutes. There is a danger of reduced direct input into cases because of the demands placed on professionals' time by the requirements of the system. There is also the danger of routinized managerial responses taking over from professional on-the-spot assessment of situations. To what extent these problems and dangers exist is not fully known and remains an issue for research. There is a pressing need for more systematic evaluation of child protection work in Britain. Steps should be taken to analyse the process of cases into and through the system and there needs to be proper evaluation of case outcomes. As was mentioned in Chapter 8, there is also very little research into consumer views, those of mothers, fathers and children. We need to know more about what professionals working within the child protection system think of it. Currently available evidence is impressionistic.

Social workers who have a key function in the system seem to have mixed feelings about the way in which it operates. On the one hand, they would like to work with a greater degree of discretion. On the other, they welcome the support and back-up the system provides, particularly in the light of criticism of individual workers for their mistakes in the past (see Corby 1987). There is little evidence about other professionals' views. Medical doctors in Britain have been critical of case conferences for being over-bureaucratic, long-winded and time-consuming. There is very little information on a police perspective or that of health, school and probation personnel.

Other important questions requiring answers are: what demands are likely to be made of the system in the future, what resources will be needed to meet these demands and how can they best be deployed? The immediate future of the child protection system is laid out in *Working Together under the Children Act 1989* (DoH 1991b). Little change is envisaged in terms of the way the system will operate. The main function of this document is to tie up the procedures to fit in with the requirements of the new act and to incorporate lessons learned from mistakes in investigations into organized abuse. This will probably prove inadequate to the problems of child protection in the 1990s.

A broad research-based review of the whole child protection system, with questions such as those raised above as its focus, is needed to assess the overall impact of child protection work. Such a review would provide a more solid foundation for future developments in this field.

Summing Up

We had in mind the notion of providing such a knowledge base for the relatively autonomous practitioner working within the child protection system. It seemed that current practice in this field was more concerned with systems of managing child abuse and, to a lesser extent, intervention methods than with applying knowledge derived from research, and that the knowledge base being used in practice was extremely limited. Some of the knowledge areas covered do not seem to have immediate application. Knowledge of the response to child mistreatment through the ages, the politics of the development of the current child protection system, the problems associated with defining abuse and the extent of the problem do not provide immediate answers to pressing problems of practice. However, we hold to the value belief that practitioners who are aware of the moral and political parameters of child protection work are likely to prove more effective (provided they have practice skills as well) than those who do not have this perspective. Similarly, a broad theoretical understanding of why abuse happens is likely to enhance child protection practice.

This of course remains to be empirically tested and flies in the face of in vogue views that demonstration of competence in practice is the sole issue. Other knowledge areas reviewed (those seeking to find out who abuses whom, and with what consequences) seem to be more closely related to the concerns of practice. While much of this research appears to have direct relevance, we have tried to point out some of the weaknesses and the problems associated with applying it to practice as well as its strengths.

APPENDIX: CHILD PROTECTION CHECKLISTS

Greenland's (1987:185) high-risk checklist includes the following items:

For Parents

1. Previously abused/neglected as a child.
2. Age 20 years or less at the birth of first child.
3. Single-parent/separated; partner not biological parent.
4. History of abuse/neglect or deprivation.
5. Socially isolated—frequent moves—poor housing.
6. Poverty—unemployed/unskilled worker; inadequate education.
7. Abuses alcohol and/or drugs.
8. History of criminally assaultive behaviour and/or suicide attempts.
9. Pregnant—post partum—or chronic illness.

For Children

1. Was previously abused or neglected.
2. Under 5 years of age at the time of abuse or neglect.
3. Premature or low birth-weight.
4. Now underweight.
5. Birth defect—chronic illness—developmental lag.
6. Prolonged separation from mother.
7. Cries frequently—difficult to comfort.
8. Difficulties in feeding and elimination.
9. Adopted, foster- or step-child.

Browne and Saqi's (1988:68) checklist includes the following items:

1. Parent indifferent, intolerant or overanxious towards child.
2. History of family violence.

3. Socio-economic problems such as unemployment.
4. Infant premature, low birth weight.

Concluding Comments

In the introduction to this book, my stated aim was to provide tor those involved in working in the child protection field a critical review of as broad a child abuse knowledge base as possible. I had in mind the notion of providing such a knowledge base for the relatively autonomous practitioner working within the child protection system. It seemed to me that current practice in this field was more concerned with systems of managing child abuse and, to a lesser extent, intervention methods than with applying knowledge derived from research, and that the knowledge base being used in practice was extremely limited.

Some of the knowledge areas covered do not seem to have immediate application. Knowledge of the response to child mistreatment through the ages, the politics of the development of the current child protection system, the problems associated with defining abuse and the extent of the problem do not provide immediate answers to pressing problems of practice. However, I hold to the value belief that practitioners who are aware of the moral and political parameters of child protection work are likely to prove more effective (provided they have practice skills as well) than those who do not have this perspective. Similarly, a broad theoretical understanding of why abuse happens is likely to enhance child protection practice. This of course remains to be empirically tested and flies in the face of in vogue views that demonstration of competence in practice is the sole issue.

Other knowledge areas reviewed (those seeking to find out who abuses whom, and with what consequences) seem to be more closely related to the concerns of practice. While much of this research appears to have direct relevance, I have tried to point out some of the weaknesses and the problems associated with applying it to practice as well as its strengths.

In this and the previous chapter, attention has focused on practice research and policy issues. The result may well be that what on the face of it seems a simple issue, the mistreatment of children, comes across as a complex and at times confusing problem without the possibility of easy resolution. If that is the case, then I offer no apologies. My belief is that an attempt to understand these complexities is a necessary requirement for effective child protection work.

APPENDIX: CHILD PROTECTION CHECKLISTS

Greenland's (1987:185) high-risk checklist includes the following items:

For Parents

1. Previously abused/neglected as a child.
2. Age 20 years or less at the birth of first child.
3. Single-parent/separated; partner not biological parent.
4. History of abuse/neglect or deprivation.
5. Socially isolated—frequent moves—poor housing.
6. Poverty—unemployed/unskilled worker; inadequate education.
7. Abuses alcohol and/or drugs.
8. History of criminally assaultive behaviour and/or suicide attempts.
9. Pregnant—post partum—or chronic illness.

For Children

1. Was previously abused or neglected.
2. Under 5 years of age at the time of abuse or neglect.
3. Premature or low birth-weight.
4. Now underweight.
5. Birth defect—chronic illness—developmental lag.

6. Prolonged separation from mother.
7. Cries frequently—difficult to comfort.
8. Difficulties in feeding and elimination.
9. Adopted, foster- or step-child.

Browne and Saqi's (1988:68) checklist includes the following items:

1. Parent indifferent, intolerant or overanxious towards child.
2. History of family violence.
3. Socio-economic problems such as unemployment.
4. Infant premature, low birth weight.
5. Parent abused or neglected as a child.
6. Stepparent or cohabitee present.
7. Single or sep arated parent.
8. Mother less than 21 years old at the time of birth.
9. History of mental illness, drug or alcohol addiction.
10. Infant separated from mother for greater than 24 hours post-delivery.
11. Infant mentally or physically handicapped.
12. Less than 18 months between birth of children.

It is interesting to consider these items in the light of the research findingds set out in Chapters.

NOTES

2. Childhood, Child Abuse and History

1. O'Hagan (1989: 21) vehemently defends a position that ignores history: 'such erudite learning, is, quite frankly, irrelevant'.
2. There are some British histories of social work. Young husband (1978) deals with the period from 1950 to 1975. Yelloly (1980) looks at the influence of psychoanalysis on social work practice from the 1930s onwards. The most comprehensive studies

are those by Woodroofe (1962) and Seed (1973).

3. See Ferguson (1990 121-2) for a discussion of this issue.
4. The best example of this kind of approach in the child protection field is the work of Linda Gordon (1989), which will be considered in more detail in Chapter 3.
5. Mackie and Taylor (1990:198-9, 402-3) provide some useful tables outlining the legal rights and responsibilities of young people at the ages of 16 and 18.
6. Frost and Stein (1989:10-16) provide a good thoughtful critique of Aries's work.
7. See Hunt (1970), Hoyles (1979), Thane (1981), Pollock (1983), Macfarlane (1979), Boswell (1988), Wilson (1984) and Hanawalt (1977).
8. See Shorter (1976), Stone (1977), Plumb (1975), De Mause (1976) and Badinter (1981).
9. See Macfarlane (1970: 203): This day a quarter past two in the afternoon my Mary fell asleepe in the Lord, her soule past into that rest where the body of Jesus, and the soules of the saints are, shee was: 8 years and 45 dayes old when shee dyed, my soule had aboundant cause to blesse god for her, who was our first fruites, and those god would have offered to him, and this I freely resigned up to him (,) it was a pretious child, a bundle of myrrhe, a bundle of sweetnes, shee was a child of ten thousand, full of wisedome, woman-like gravity, knowledge, sweet expressions of god, apt in her learning,] tender hearted and loving, an [obed]ient child [to us.] it was free from [the rudenesse of] little children, it was to us as a boxe of sweet ointment, which now its broken smells more deliciously than it did before, Lord I rejoyce I had such a present for thee, it was patient in the sicknesse, thankefull to admiration; it lived desired

and dyed lamented, thy memory is and wille bee sweete unto mee' (26 May 1650).

10. Davin's account (1990) of the way in which this combination of factors led to coercive, class-blind philanthropy is particularly interesting.
11. It should be noted that Boswell has been criticised by some historians for being over-optimistic about the fate of abandoned children (see Tilly *et al.* 1992).
12. This statement, while demonstrating public concern about child sexual abuse, also shows that there was probably a hierarchy of concerns. No mention is made of female children and those of non citizen status.
13. Lafontaine (1990: 210) drawing from anthropological work, lends some support to such a view with regard to sexual abuse.

3. A Recent Political History of Child Abuse and Neglect

1. The campaigns created by Kempe in the USA (see later) and the fund-raising practices of the NSPCC are examples of this process.
2. See Dingwall *et al.* (1984: 214).
3. See Rose (1986: Chapter 11).
4. It would be wrong, however, to suggest that there was no relevant legislation at all. The 1868 Poor Law Act made wilful neglect by a parent of a child under 14 that threatened or resulted in serious injury an offence. The weakness of this law was that only Poor Law officials were empowered to bring cases of this kind to court and they very rarely did so. The 1861 Offences Against the Person Act could also be used to prosecute parents for assaults on their children, but no particular agency was mandated to report or seek out such abuse.

5. Behlmer (1982: 172) shows that 58 per cent of the 23,124 cases reported to the NSPCC in 1896-7 emanated from the general public.
6. Between 1908 and 1920, the number of recorded cases of incest rose from a yearly average of 24 to 57. This average gradually increased to 101 by the beginning of the Second World War. By 1950, the average had increased to 237. Between then and 1985 the annual number of cases fluctuated between a low of 230 to a high of 334. The annual figures for 1985, 1986 and 1987 show a marked increase: 444, 511 and 516 (Home Office Criminal Statistics).
7. The major legislative change of this period was the 1933 Children and Young Persons Act, which followed in the tradition of previous legislation by extending the range of prosecutable offences against children. It also blurred distinctions between neglected children and young offenders by such measures as the creation of approved schools for all who came into these two categories (before this they had attended separate institutions). By 1946, when the Curtis Committee reported, this type of thinking was well established: 'According to the evidence of the Home Office, it is often an accident whether a child is before the court for an offence or as a neglected child, and it is accordingly appropriate that the same methods of treatment be equally available.'
8. See pp. 209-10, which provide an extract of an NSPCC inspector's discovery of and response to a child sexual abuse case. The case is undated, but as the children involved were committed to industrial schools, it must have been before the implementation of the 1933 Children and Young Persons Act. The case involved a 12-year-old girl, her sister, 11, and her brother, 9. Their mother was dead and they were living with their father and a 46-year-old

male lodger. It was revealed that the two girls slept in the same bed as the lodger. The inspector's records described the children as being dirty and neglected. A medical examination was arranged by the inspector. 'This proved that the girl aged 12 had been interfered with.' The father and the lodger were prosecuted. The father, a first-time offender, received a light sentence, the lodger 18 months' hard labour. The children, as mentioned above, were committed to industrial schools. The tenor of the report is very factual. The response to this case comes across as insensitive by our standards but unequivocally child protective.

9. This does not mean that the approach at the ground level was not authoritative and intrusive (see Alien and Morton 1961: Chapter 6).
10. See Greenland (1958) and Alien and Morton (1961: 89-93).
11. Nelson (1984) charts the development of child abuse as a social problem in the USA and shows how, from a topic that hardly raised a flicker of concern in the 1950s, it came to figure as a prominent part of Walter Mondale's presidential campaign in 1974.
12. Parton (1985:61) points out that the unit produced 17 articles in professional journals between 1969 and 1973.
13. The number of place of safety orders doubled in the period after the publication of the Maria Colwell report. In March 1974 there were 353 children in care on place of safety orders. In March 1975 there were 596 and in March 1976 there were 759. Between 1977 and 1985, the annual number of place of safety orders remained at roughly the same level, ranging between a low of 5207 per year to a high of 6613. However, Dingwall and Eekelaar (1984) argue that the numbers of children committed to care as a result of abuse and neglect did not increase

between 1970 and 1985. Also there was no major increase in the prosecution of parents for offences against children during this time (see Martin 1978: 209).

14. Cyril Greenland's testimony was influential in this respect. His views are elaborated in his book *Preventing CAN Deaths* (1987) and are further considered in Chapters 5 and 8. See Appendix.
15. See note 13 for the annual number of place of safety orders between 1977 and 1985. In 1986, they rose to 7191 and in 1987 to 8055.
16. See Armstrong (1978), Brady (1979), Angelou (1983). Bagley and King (1990:1-24) give a good overview of the public impact of personal accounts of sexual abuse.
17. See Cleveland (1988: para. 4.80, 64). A ten-year-old girl and her two siblings were examined with their mother's consent by Dr Higgs in the hospital. Their father removed them. They were subsequently brought into police custody under place of safety orders and examined by a police surgeon, Dr Beeby. His diagnosis conflicted with that of Dr Higgs, who re-examined the children. The following day they were re-examined by Dr Irvine, another police surgeon. The report adds that the children 'were later examined by three more doctors'.
18. Children Act 1989, section 44(7).
19. Children Act 1989, schedule 2 para. 5.
20. Lowe (1989) demonstrates this rate of increase. In 1971 there were 622 originating summons. By 1981 there were 1903 and in 1987 there were 3605. The bulk of these increases is accounted for by local authority applications.
21. For the Nottingham case see *The Times,* 30 November 1990. For Rochdale see *The Times, 9* August 1990, and leader article, 15 September 1990. For the Orkneys see *The Times,* 5 January 1991.

4. Defining and Measuring the Extent of Child Abuse

1. Stephanie Fox (Fox 1990) was known by the health and welfare authorities to have suffered minor bruising on 30 occasions before she died as a result of a violent outburst by her father. They had previously declined to remove Stephanie, probably because their expectations of the family were very low and it was felt that such bruising (largely perceived to be the result of careless supervision of the child rather than of physical mistreatment) was not remarkable given the background of the Fox family. The Lester Chapman case (Chapman 1979) shows a similar rather resigned acceptance of low standards of care.

   In Tyra Henry's case (Henry 1987), the social worker made assumptions about Caribbean culture that placed heavy expectations on this child's grandmother to provide care and protection for her. West Indian grandmothers were stereotypically considered to be the linch-pins of the family as far as child care was concerned. In Beatrice Henry's case, nothing could have been further from the truth. She had experienced the death of her husband and the severe mistreatment of her grandson. She was a lone parent dealing with the problems of her own three children, was inadequately housed and had multiple debts. As events proved, she was completely unable to protect Tyra.
2. There is something of a contradiction here in that, despite the general phrasing used to define child abuse in legal proceedings, lawyers are concerned that there should be clear and definite proof of abuse in court.
3. The following two vignettes, with consequences in parentheses, are taken from Giovannoni and Becerra's study and give a flavour of the general

approach used: 'The parents regularly left their child alone outside the house during the day until almost dark (neighbours have spotted the child wandering five blocks from home).' 'The parents banged the child against the wall while shaking him by the shoulders (the child suffered a concussion).'

4. Section 31(2) of the 1989 Children Act states: 'A court may only make a care order or supervision order if it is satisfied -
   (a) that the child concerned is suffering, or is likely to suffer, significant harm; and
   (b) that the harm, or likelihood of harm, is attributable to -
      (i) the care given to the child, or likely to be given to him if the order were not made, not being what it would be reasonable to expect a parent to give to him; or
      (ii) the child's being beyond parental control.'
      Section 31(9) states: '"harm" means ill-treatment or the impairment of health or development; "development" means physical, intellectual, emotional, social or behavioural development; "ill-treatment" includes sexual abuse and forms of ill-treatment which are not physical.'

5. This failure to give an explanation of the demise of the 'grave concern' category is most surprising considering that it accounts for half the children on child protection registers. The provisional feedback to local authorities on register returns for the year ending 31 March 1991 merely states that the category will not be used for registrations after 14 October 1991 (the date of the implementation of the Children Act). While this move is to be welcomed, as it will force professionals to be more specific about their concerns, some explanation of

the reasons for taking this course of action would also have been welcome.

6. See Baldwin and Oliver (1975), Smith (1975) and Greenland (1987). However, these are all retrospective studies. Although the vast majority of the severely ill-treated children they studied had minor bruising at an earlier date, this does not mean that all children with minor bruising will be severely abused in the future.
7. See section 4.6 pp. 12-13.
8. Issues of this kind were highlighted in the Doreen Aston report (Aston 1989). Before Doreen's birth, her mother Christine Mason had a child, Karl, who died aged 10 weeks. A post-mortem revealed that he had three fractured ribs and a subdural haemorrhage, but these injuries were not thought to have contributed to his death, which was finally recorded as a cot death (see p. 8, paras 11 and 12). The authors of the report felt that the cot death decision led social workers and others to overlook the fact that the child had been abused and, therefore, that subsequent children might be particularly at risk.
9. Polansky and colleagues argue in later works that there are common agreed standards of care. They developed what they termed a 'childhood level of living scale' (Polansky *et al.* 1978) and found that there was considerable agreement among lay people about what was neglectful (Polansky and Williams 1978). Polansky *et al.* (1983: 345) conclude from these previous studies that 'it appears there is such a thing as an American standard of minimal child care that is commonly held and that may be invoked in the definition of child neglect for legal and social work purposes.' Such ideas do not, however, seem to have influenced practice greatly.
10. Political factors may also play a part in this relative

lack of urgency. Wolock and Horowitz (1984) argue that neglect of children is far more common than physical abuse in the USA, but receives far less attention. This 'neglect of neglect', as they term it, results from the fact that closer inspection of many children's lives would reveal the extent to which poverty contributes to neglect and this would create political embarrassments for governments.

11. Pages 69-74 of the Beckford inquiry report provide useful detailed material on the failure-to-thrive syndrome. At birth, Jasmine weighed 5 Ib 11 oz. By four months, she had reached an average weight for a child of her age and, therefore, would have been expected to maintain this average growth throughout her early childhood. After ten months, Jasmine had slipped back and her weight was well below the average. It was below the third centile (i.e. Jasmine was among the 3 per cent most poorly developed of all children). At age 15 months Jasmine weighed 18 Ib 6 oz, even further down in the third centile. At 20 months, when she experienced her first serious injury, she was still well down in the third centile. At the age of 27 months, after she had been in foster care for seven months, Jasmine had grown considerably and weighed 25 Ib 5 oz (on the 25th centile). At this point she was returned to her mother and stepfather. She was not weighed again until her death 27 months later. She weighed 23 Ib.

12. There is much current concern about the perceived extent of sexual abuse of children by adolescents, (see Chapter 5). There is some uncertainty as to whether such offenders should be seen as abusers or whether they should be responded to as in need of help themselves. Are such offenders sexually mature?

13. See Finkelhor *et al.* (1986: 66).

14. It is important to distinguish between the terms 'prevalence' and 'incidence'. Prevalence studies measure how many people in a given sample have experienced a particular phenomenon at least once over a particular period of time. Incidence studies measure the number of occurrences of a particular phenomenon in a given sample of people over a particular period of time. Thus, if a sample of 100 people were asked if they had been sexually abused at least once before the age of 15, and 20 said that they had, the prevalence rate of abuse of such children would be 20 per cent. If these 100 people were asked how often they had been sexually abused before the age of 15, the answer would probably be higher. They might report abuse on 60 occasions. The incidence rate of abuse would then be 60 per 100 people over the first 15 years of their lives. Lafontaine (1990: Ch. 2) gives a fuller, very useful account of these and other issues relating to prevalence and incidence studies.
15. Those families with just one- and two-year-old children were omitted because one of the study's objectives was to measure sibling violence. How this affects the interpretation of the findings (i.e. are they an under-or over-estimate of the total) is unclear. In 1990-1, the registration rate for children physically abused in the first year of their lives was higher than for all the other age groupings.
16. It must be remembered that this rate covers 15 years of childhood but does not take into account the number of abuse incidents that a child may have experienced. The MORI Poll survey found that 63 per cent of those abused had been abused once, 23 per cent were subject to multiple abuse by the same person and 14 per cent subject to multiple abuse by a number of people. Estimating accurately the gap between officially responded to cases and

the true number is, therefore, a highly problematic endeavour.

5. Who Abuses Whom

1. In the case of Wayne Brewer, who was killed by his stepfather, Nigel Briffett, his mother was heavily implicated. In a report written for court when Wayne was made the subject of a care order, the social worker commented on her as follows: 'Her inability to restrain her husband, together with the rather negative handling of the child, characterised by her unwillingness to readily handle him, and to generally care for and stimulate him, indicate that she has not really been able to accept responsibility for him' (Brewer 1977).
2. The views put forward on this subject in the Cleveland report are of interest. The term 'collusive mother' is not explicitly used. However, the implications are clear. 'Again quoting Professor Sir Martin Roth, "In many cases mothers play a role in the genesis of the sexual abuse of their daughters. They may be too physically ill or inadequate in personality to provide proper care and protection for their children. In other cases mothers elect the eldest cr one of the oldest daughters to the role of 'child mother'. The girl in her early teens or even earlier is expected to take the responsibility for the caring of the younger children whose mothering role is allowed to slide into a sexual relationship with the father. This is tolerated with little or no protest. I refer to lack of protest of [sic] the part of the mother for a variety of reasons and the mother may in such cases deny what is happening. She conceals the truth from herself as well as others; the relationship continues and when the situation is brought to light it may be insisted by the mother that it had been unknown to her"' (Cleveland 1988: para. 29, 8).

3. Mattinson and Sinclair (1979) point to the lack of involvement of males by social workers in marital work! Corby (1987: 91) found that little work was done with fathers or male parent substitutes in child protection work. The Stephanie Fox inquiry commented on the fact that very little was known about her natural father, Stephen Fox. 'So far as Stephen Fox is concerned, he was, and is, as we shall see, a shadowy figure. We believe that it would have been very useful to know more of his wider family relationships, to bring him out of the shadows. This is not to suggest that pointers would necessarily have emerged towards the final appalling violence. We do, however, believe that if, as surely must be the case, the social workers wanted to facilitate change in the behaviour of these parents, they had to understand more about it' (Fox 1990: para 4.9, 24). Cultural factors account for part of this pattern of intervention in that males (particularly in the lower classes) are not normally expected to be closely involved in child care matters. In female-headed lone-parent families, men's involvement with children and family matters is probably even more peripheral. However, there does seem to be a tendency for social workers and other professionals (e.g. health visitors) to fall too readily into working exclusively with females in families because of assumptions about their mothering role and also because of their availability.
4. That women do physically abuse children is not being denied. For a useful case study of women who have fatally abused their children see Korbin (1989).
5. Nevertheless, Greenland (1987:185) incorporates the category 'age 20 or less at birth of first child' in his high-risk checklist (see Appendix).
6. This trend may be being reversed somewhat by

anti-racist training and policy implementation. However, these innovations are somewhat patchy and vary considerably from region to region.

7. This is a problematic issue because of the degree of secrecy that shrouds intra-familial child sexual abuse and the fact that we cannot know for certain whether or not someone who appears to have focused his abuse on one child has offended against other children or whether he might do so in the future.
8. In 1989, 26 per cent of all live births were outside marriage. In 1987 lone-parent families constituted 14 per cent of all families with dependent children (over 900,000 in all). However, although the numbers of such families rose dramatically throughout the 1970s, they have remained stable throughout the 1980s. In 1989, there were 153,000 divorces (decrees absolute) involving 148,000 children under the age of 16. Again these figures have remained stable throughout the 1980s after a steep increase in the previous decade. However, the number of under-four year olds involved in divorces has risen by 20 per cent in the 1980s (figures from the OPCS General Household Survey 1987, and from OPCS marriage statistics for 1989).
9. Christine Mason, the mother of Doreen Aston, was known to have been depressed in October 1985, seven months before Doreen was born. In January 1985, her ten-week-old son, Karl, had died in suspicious circumstances (see Chapter 4, note 8). Nine months later Christine was said to be continuing 'to give cause for concern as she was carrying the ashes of her dead child around with her' (Aston 1989: Chapter 2, para. 15, 9). There is no further reference to depression or grief reaction in the report.

   Beatrice Henry, Tyra Henry's grandmother, on whom great reliance was placed by Lambeth Social

Services Department for her protection, had experienced the death of her husband and the maiming and loss of her grandson in 1982 (she had also previously suffered the death of her own son). The inquiry report points out that there was 'not a line in the contemporary records and not a line in the evidence given to us which recognises that by the time of Tyra's birth, Beatrice Henry was struggling with private grief along with the difficulties of her daily life' (Henry 1987:112). The inquiry panel recommended that social workers receive training in this area and be directed to devote attention to such matters in future.

The Koseda inquiry (Koseda 1986) noted that the cohabitation of Heidi Koseda's mother, Rosemary, with Nicholas Price 'seems to have been the start of a marked deterioration in her mental state and way of life, which culminated in serious mental illness after the discovery of Heidi's body early in 1985' (para. 1.1).

10. These other factors (except for marital or partner problems) are all included in Greenland's high-risk checklist for parents (see Appendix).
11. In the cases of Tyra Henry (Henry 1987) and Darryn Clarke (Clarke 1979) there do not seem to have been obvious signs of conflict. However, very little information seems to come to light about the relationships of parents or parent substitutes in public inquiry reports. Both Andrew Neil and Charles Courtney, who were convicted for the killing of these children, had histories of violence. The lack of conflict in their relationships with these children's mothers may have been the result of their being totally dominating and controlling figures.
12. See page 122 of the Beckford report for comments about lack of attention paid by welfare workers to the stresses of pregnancy.

13. See Dingwall (1989). These issues will be further discussed in Chapter 8.
14. Doreen Aston, Stephanie Fox and Tyra Henry and their families were all under financial and material stress at the times of their births and in their early childhoods. This was particularly apparent in the case of Stephanie Fox. The Fox family, consisting of mother, father, Stephanie aged two, and twins, aged eight months, were eventually provided with accommodation on the nineteenth floor of a tower block. They suffered chronic financial problems, despite help being given by social workers under section 1 of the 1980 Child Care Act. The report, while acknowledging this help, comments: 'Despite the desperate circumstances of families in crisis of this kind, we believe that section 1 financial help should be used whenever possible as part of a consistent plan to promote the well-being of children in families, rather than as a routine response to crisis calls for help' (Fox 1990: 72). It remains to be seen whether section 17 and Schedule 2 of the 1989 Children Act, which deal with material aid, will be implemented in this constructive way.
15. Stephen Menheniott was 19 years old when he died as a result of multiple injuries inflicted him over a long period of time by his father. Because of his age, this case could be seen to be the murder of a young adult, but this judgement would belie the true nature of Stephen's abuse. The events of his life and death all carry the hallmarks of a case of child abuse. His father, Thomas Menheniott, was brought up in public care institutions. Nearly all his eight children had long spells in care. He had convictions for neglect and ill-treatment of his children and had been acquitted after being charged with incest. Stephen was a rather pathetic, immature young man. He had spent the bulk of his life in

residential care attending schools for the maladjusted and had developed few lasting relationships. He was returned to his father's 'care' at the age of 15. He was clearly exposed to prolonged physical abuse and intimidation.

6. The Causation of Child Abuse

1. There has been increasing emphasis throughout the 1980s on the 'management' of child abuse by developing systems to ensure that cases are properly identified, registered and monitored. There has been a concomitant shift away from trying to assess the reasons for abuse happening and using this assessment as the basis for the treatment of the problem (see Corby 1987; Home 1990). This has been a direct result of the heavy criticism that professionals in the child protection field (especially social workers) have been subjected to from public inquiries into child deaths by abuse. It also reflects the influence of systems-based managerial approaches, which have gained more and more acceptance generally in the welfare field (Davies 1981).
2. Sociobiology emerged as an identifiable discipline in the 1970s (Wilson 1975). It is essentially a new form of social Darwinism, but is more subtly argued than its predecessor. Sociobiologists' main proposition is that social sciences have been too narrow in their interpretations of human behaviour, associating it almost exclusively with cultural and social influences. In this process, the fact that humans are, like any other form of living species, biologically driven, has been forgotten. From a sociobiological perspective, biological forces, particularly gene preservation, are paramount influences on behaviour (Dawkins 1976). Social scientists have responded by arguing that

sociobiology has theoretical flaws and adverse political effects, such as the potential for encouraging racism, sexism and far right views such as eugenics (Sahlins 1977; Montagu 1980). In many ways, this conflict is part of the continuing nature-nurture debate which has existed formally since the emergence of the social sciences in the eighteenth century.

3. For a mainstream analysis see Stafford-Clark (1965), Fancher (1973) and Jahoda (1977). Kline (1981) provides details of empirical studies into Freudian theory. Feminist critiques of Freudian theory are to be found in Mitchell (1974: 61-108) and Sydie (1987: 125-67).

4. Steele and Pollock (1974:128-30), in discussing the non-abusing parent, point out that The other parent almost invariably contributes, however, to the abusive behaviour either by openly accepting it or by more subtly abetting it, consciously or unconsciously.' They list various forms in which this happens, one of which seems to have been apparent in several of the inquiry report cases in Britain (e.g. Darryn Clarke, Jasmine Beckford, Kimberley Carlile): 'One parent feeling overwhelmed and frustrated may turn the infant over to the other with admonitions to do something more drastic to stop the baby's annoying behaviour.' How voluntary such behaviour is, is open to question. The non-abusing parent may be forced to hand over the child for such inappropriate discipline. With regard to the contribution of children to their own abuse they stress that 'characteristics presented by the infant, such as sex, time of birth, health status, and behavior are factors in instigating child abuse.'

5. Steele and Pollock's views on the influence of socio-economic factors on the incidence of child abuse are that they are marginal: 'Basically they are

irrelevant to the actual act of child beating. Unquestionably, social and economic difficulties and disasters put added stress on people's lives and contribute to behavior which might otherwise remain dormant. But such factors must be considered as incidental enhancers rather than necessary and sufficient causes' (p. 108).

6. One exception is a study of intervention into failure to thrive cases (Iwaniec *et al.* 1985).
7. Corporal punishment has been outlawed in schools in Sweden since 1952 and legislation was passed in 1979 prohibiting the use of physical force on children within the family (see Gelles and Edfeldt 1986).
8. Smith (1991) outlines the parts of the Children Act that specifically enhance a children's rights perspective. Two examples are: (a) section 22, which emphasizes the need to consult children of sufficient age and understanding with regard to all decisions that affect them (this section is very similar to section 18 of the 1980 Child Care Act); (b) section 44(7), which gives a child of sufficient age and understanding the right to refuse a medical examination under an emergency protection order (a direct result of events in Cleveland, where it was felt that children were exposed to many assessments and examinations without having any recourse if they objected). In general, the 1989 Children Act is a move towards greater consideration of children as subjects rather than objects. However, the legislation still leaves a lot of room for adultist views to prevail.
9. Lynch (1976) provides a useful integrative model called 'critical path analysis'. This incorporates the first three of Belsky's systems, but not the macrosystem, which takes into account broader societal issues.

7. The Consequences of Child Abuse

1. See also Zimrin (1986) and Rutter (1985).
2. Death is also a consequence of child abuse cases. The extent of death by child abuse as reported in official statistics is discussed in Chapter 4.
3. Transmission of abuse from one generation to another is also an important (though highly contentious) long-term effect. This has already been dealt with in Chapters.
4. Green (1978) found significantly higher rates of self-abusive behaviour among a sample of physically abused children compared with neglected and non-abused children.
5. Jasmine Beckford's stepfather, Morris, spent the first nine years of his life with his grandmother in Jamaica before being reunited with his parents in Britain. At age 13, he and his sister were severely beaten by both parents and forced to sleep in an outhouse with no bed and one blanket between them before being taken into care (Beckford 1985: 42). It is not known how Tyra Henry's father, Andrew Neil, was treated as a child. His mother left home when he was aged seven. He was noted to have hysterical fits at age ten and physically assaulted a two-year-old baby when he was aged 13 (Henry 1987: 8-10). Unfortunately, not enough is known about the background of fathers in many other public inquiry cases for a variety of reasons (see Chapter *5).* Closer attention to their biographical details (and those of the mothers) would undoubtedly add to our understanding of the causes of such gross violence to children.
6. Carmen *et al.* (1984: 382) show considerable awareness of the potential impact of physical and sexual abuse on women: In our sample, the abused females directed their hatred and aggression against themselves in both overt and covert ways. These

behaviours formed a continuum from quiet resignation and depression to repeated episodes of self-mutilation and suicide attempts. . . . Markedly impaired self-esteem was prominent among these patients, as they conveyed a sense that they were undeserving of any empathic understanding or help by clinicians.' These reactions were in contrast to the outwardly directed aggression that was most common among the mentally ill males in this study with a history of abuse. From this perspective, the form of the reaction to abuse is heavily influenced by social and socialization factors. Similarly, being black and abused in predominantly white-dominated societies also influences the consequences and the response to the abuse (see Angelou 1984).

7. There is still a major problem in deciding what is or is not appropriate or excessive sexual behaviour, particularly with regard to older children.
8. Browne and Finkelhor (1986) also note that children who experience abuse by more than one person are more likely to experience harmful effects.
9. Gordon (1989) points out that some of what have been termed 'negative' effects of child sexual abuse, such as running away and inappropriate sexual behaviour, could be seen as positive reactions to the situations many of the girls in the records she studied were faced with. '"Sex delinquency" was an escape route not only of victims, but often of highly responsible victims, trying to avoid telling their secrets and exploding their families' (p. 240).

8. Research into Child Protection Practice

1. The Doreen Aston report is particularly useful for developing an understanding of the role of the health visitor in child protection work and the pressures created in working in difficult environments with limited resources.

2. Porter (1984) provides a checklist of factors linked with child sexual abuse, as do Finkelhor and Baron (1986). These do not seem so far to have been applied to child protection practice in the way that has been the case with regard to physical abuse and neglect. This could be explained by the now widely accepted view that action should only be taken in cases of intra-familial abuse when there is clear evidence.
3. The contrast with the situation in the field of physical abuse is ironic. In the Beckford report social workers were criticized because of their lack of knowledge about relevant research and the findings of previous inquiry reports. In Cleveland, they were rebuked for applying knowledge derived from research in an uncritical fashion.
4. Darryn Clarke was never seen by social workers before his death. The focus of the inquiry was on the response to the initial allegation and the reasons for the delay in tracing a child who was said to be at risk. Similarly, Kimberley Carlile was not known to the social services department to whom allegations of ill-treatment were made. However, in this case there had been prior social work involvement by other social services departments. The inquiry concentrated on the delays and difficulties associated with assessing whether Kimberley was at risk or not.
5. Parents Anonymous groups were first set up in California in 1969 (see Holmes 1978). Small voluntary groups of parents who have abused their children meet a professional group helper to share and discuss feelings and issues. The aim is to help such parents improve their self-image and provide support for each other. Parents are encouraged to contact each other regularly outside the group and to follow up parents who drop out. There is no

direct equivalent in Britain. Organizations such as Parents Against Injustice and the Family Rights Group are pressure groups organized to challenge the child protection system.

6. See Ong (1985) for missed opportunities in an NSPCC day/nursery centre.
7. See Schedule 3, sections 2 and 3 of the 1989 Children Act. These sections deal with the powers of supervision orders. The court has the power to require the subject of a supervision order to attend a specified place (e.g. a day nursery) under the directions of the supervisor. Provided he, she or they consent, the responsible person(s) can also be required to attend a specified place under the directions of the supervisor (e.g. a family day or residential centre). Failure to comply with these requirements can result in the supervisor returning to court to seek a variation in the order (section 39).

9. Current Issues in Child Protection Work

1. The British system of intervention into child abuse, particularly sexual abuse, is considered to be more authoritative and intrusive than in other European countries where, as a rule, there seems to be greater emphasis on containing problems within the family and less involvement of the criminal justice system (see Bunting 1991). In countries such as Belgium, Holland and Germany, for instance, the police have a much lower profile in such work than is true in Britain. Supporters of the British system argue that to tackle child abuse properly it must be seen as a crime and they consider the continental system to be behind the times. On the other hand, advocates for the continental approach see it as more flexible and advanced, and likely to achieve better results than the more punitive British method.

2. It is difficult to understand the intentions of the *Working Together* guidelines in this respect. Virtually all the examples used in these guidelines seem to have the case of sexual abuse in mind. For instance: 'Awareness of the needs of the child should focus the enquiry on the child. Every effort should be made to help him or her to relax and feel at ease. Consideration should be given to a child having a parent, relative, friend or supporter present during the investigative interviews, as the circumstances of the case determine. . . . The interviewer must listen carefully to what the child has to say and communication with him or her must be in a responsive and receptive manner. He or she must work at the child's pace and use language that the child can understand and thus enable the child to talk about and give as clear an account as possible of events that have taken place' (DoH 1991b: 5.14.7). An interview of this kind has not been typical practice in the case of physical abuse. This is not to argue that it is not a good idea. However, if it is the intention of the guidelines to encourage similar approaches for physical and sexual abuse investigations, then it should state its objective unequivocally.
3. Gordon's (1989) study of the recent history of intervention into child abuse in Boston in the USA charts the link between the existence of strong feminist movements and greater attention being paid to abuse of children and women in families: 'For most of the 110 years of this history, it was the women's-rights movement that was most influential in confronting, publicizing, and demanding action against family violence. Concern with family violence usually grew when feminism was strong and ebbed when feminism was weak' (Gordon 1988:4).

4. The last resort principle is not legally defined as such, but section 1(5) of the 1989 Children Act states that the court 'shall not make any order or any of the orders unless it considers that doing so would be better for the child than making no order at all.' Under the previous legislation (the 1969 Children and Young Persons Act) the court had a similar duty, namely to consider whether the making of an order was necessary to ensure sufficient care or control of a child.
5. Before the 1989 Children Act, there were three ways of taking action on the grounds of likely risk in the future: under sections 1(2)(b) and 1(2)(bb) of the 1969 Children and Young Persons Act and under wardship proceedings (section 7, Family Law Reform Act 1967).
6. In London, there were reports of 600 registered child abuse cases being unallocated in 1990. See the 1990 Social Services Inspectorate report, *Child Protection in London. Aspects of Management Arrangements in Social Services Departments.* London: HMSO.

6

UN Action and National Policies for Children in Developing Countries

> "The best way of training the young, is to train youself at the same time; not to admonish them, but to be seen always doing that of which you would admonish them."
>
> —*Plato*

It is pertinent to note that in 1979 the world observed the Year of the Child. The implications of this action by the UN General Assembly to focus attention on the child during 1979 are enormous. The child born in 1979 will 'come of age' and reach 21 years in the year 2000. This means that the Third millenium will directly involve as adults those children being born now. The world has 21 years to work for a better environment for its children and the young adults of the future.

Needs and Rights of the Child

The necessity of planning for the needs of children may be indicated from the following facts:

- 20 per cent of all children born in a (typical) underdeveloped country, will have died within a year of their birth;
- of the remainder many will die during the following preschool years, or will be handicapped throughout

their lives due to childhood disease or lack of adequate nutrition;

- of those who will survive until school age, many will not get to primary school and only a very small proportion will have the chance of acquiring even a rudimentary education;
- of those who leave primary school only a few will get vocational training; and the majority of adolescents will swell the ranks of the unskilled and the unemployed.

The foregoing facts highlight the needs of children in developing countries. These needs have been articulated more fully in the Declaration of Rights of the Child, unanimously adopted by the United Nations General Assembly in 1959. The Rights are to:

1. A name and nationality; and equal status for every child at birth.
2. Maternal protection and family care.
3. Health care and protection against diseases.
4. Adequate nutrition (which changes with the child's age), and adequate shelter.
5. Education.
6. Play and recreation.
7. Social protection irrespective of colour, race, language, etc.
8. Protection from work for child, the and from exploitation for youth.
9. Special care for physically, mentally or socially handicapped children, such as orphans, retarded children, etc.
10. Protection from racial, religious and other forms of discrimination.

Meeting the Needs of the Child

It is in the following areas that much of the specific

action envisaged in the interest of children and adolescents has to be concentrated:

(i) Development of services of health, maternal and child health, family planning, supply of potable water, and environmental sanitation;

(ii) improved nutrition, especially for infants and young children and pregnant and nursing mothers;

(iii) development of human resources, including provision of minimum educational opportunities for children, educational and vocational preparation of adolescent boys and girls, and expansion of opportunities for absorbing adolescents and youth entering the labour market into productive work; and

(iv) development of appropriate welfare services, community institutions and voluntary agencies to provide, in particular, for the growth and development of the young child. More generally, it will be necessary to strengthen the family and the community, to prepare citizens for responsible parenthood, and to develop skills and opportunities for gainful and productive work for girls and women.

Health

In most developing countries there is a general, and often extreme, scarcity of staff and facilities in the rural areas. The situation is only slightly better in towns. An infant death rate three or four times as high as in industrialised countries and a young child death rate 30 or 40 times as high have, as their main causes, respiratory and diarrhoeal diseases, malnutrition and vector-borne and parasitic diseases, that have largely disappeared from the industrialised countries. Moreover, as WHO has stressed, the health needs of children cannot be considered separately from the health needs of the family. There is need to link up programmes for the improvement of health and diet for the whole family with health programmes for the children.

Nutrition

WHO has said that 'adequate nutrition must be given the highest priority amongst the needs of children'. Most experts emphasise the importance of providing adequate protein especially to the young child. If development proceeds from the application of brain power, it is crucial to remember that the human brain mainly forms before the age of three years, and that this is largely a process of protein synthesis. Moreover the nutritional quality of various sources of protein must be taken into account. Children suffering from a deficiency of calories or from diseases cannot absorb or convert their protein intake. It is an unfortunate fact that the main cereals contain a smaller proportion of protein than is required for children.

TABLE 1

Protein Content of Staple Crops

Crop	*Protein Content (per cent)*
Maize	10
Wheat	10
Yam	5-10
Rice	7
Cassava	1

The pattern of consumption in some developing countries is compared with that of the United Kingdom.

Malnutrition is a 'polite word for semi-starvation among children'. It leads to the growth of diseases. Protein deficiencies, in particular, give birth to many diseases: e.g., kwashiorkor, a disease in which the child becomes swollen with oedema. Vitamin deficiencies (research has isolated Vitamin A as the main one) can lead to blindness. Rickets are caused by the absence of Vitamin D. Some other diseases caused by malnutrition are anaemia, infectious diseases and growth failure and diseases of the brain. The last mentioned are often the most dangerous as the damage is sometimes incurable and permanent.

TABLE 2

Pattern of Consumption in Developing Countries

Country	*Average daily per capita supply in grams*		
	Cereals	*Meat, Fish, etc.*	*Milk etc.*
Brazil	600	104	155
India	370	8	122
Philippines	489	25	47
U.K.	485	225	559

Clearly, malnutrition is common in developing countries. Its causes are complex and interrelated. Poverty, undoubtedly, is the main one, but dietary habits of the people also contribute to it To combat malnutrition among children we would have not only to expand the supply of nutritional (i.e. high protein) food but also overcome any customs and superstitions which may prevent the adoption of a balanced diet, often specifically related to the diet of children. In urban areas, the problem of malnutrition is often even more severe than in the villages. Bad sanitary conditions aggravate the problem, in both town and country.

Existing programmes of UNICEF for the improvement of child nutrition try to take account of the factors outlined above, and they include, among others, milk and vitamin distribution, and education concerning nutrition. Efforts should be directed towards making people—especially mothers—aware of the seriousness of the problem and show concrete ways of improving the quality of food provided for children. The protein enrichment of traditional foods, in various ways, is another useful contribution to this end.

Education

Less than half of the world's school age children are. in school. Table 3 gives the ratio of school-going children to total population.

About 20 per cent of the world population is in the age-group 5-14, but this proportion is lower in the developed and higher in the underdeveloped countries. Thus, even the gap between developed and developing region fails to show the full extent of the difference. The problems of providing education at different levels would require a separate special discussion. Briefly, in developing countries, there is a shortage of qualified teachers, school buildings, equipment, text-books, teaching materials, etc., at the primary level.

At the secondary level, besides the problems just mentioned, there is a great need to do away with antiquated methods and to revise the curriculum, particularly in science and technical education. In many developing countries secondary and technical education is not yet free. This imposes a heavy burden on parents. Moreover, education is hampered by very limited financial and budgetary resources.

TABLE 3

Ratio of School-Going Children to Total Population Region Ratio (Per Cent)

Region	*Ratio (per cent)*
Africa	7.2
North America	23.5
Latin America	13.1
Asia	10.3
Europe	15.5
USSR	16.0
World total	12.4

As such, there is a need to link up education programmes with the requirements of economic development in general. This again requires careful planning: resource allocation within the education sector must take into account the needs of economic development There is need to define educational policy, to set out the relation between the various levels and branches of the education system, in view of development requirements. Special attention must be paid to preparing young people for more productive work in the agricultural sector. Vocational training needs to be further emphasised.

Increases in enrolment, especially at secondary and higher levels, in developing countries, serve gradually to reduce the enrolment gap, but they also impose tremendous new strains on financial and absorptive capacity (for the higher levels).

Social Services

In a dynamic and developing society, children and youth are exposed to many pressures. Social welfare services can play a useful part in alleviating some of the pressures and directing them into productive channels. These social services must aim at strengthening family bonds, developing community relationships, reducing crime and juvenile delinquency and combating the undesirable effects of urbanisation and slum-living. Provision must be made for orphans and abandoned children, as well as for children who are chronically ill or crippled.

Action by the UN System

The United Nations family of organisations took children seriously from the very first days of its existence, by establishing an Emergency Fund to help children. This fund, was initially specifically to help children and ensure that they were not forgotten in the postwar chaos and the growing amounts of relief aid being provided.

The Fund was called the United Nations International Children's Emergency Fund (UNICEF) and most of its initial efforts were directed to helping children in parts of Europe and Asia most seriously affected by the war, during the second half of the 'Forties'. As Europe began to rehabilitate itself the emphasis of UNICEF increasingly shifted to assisting children in developing countries of Asia, Africa and Latin

America, especially the growing number of countries which had recently acquired independence from colonial rule.

While UNICEF led the way, other agencies and programmes of the UN system became increasingly involved in assistance to the developing countries, first among them the World Bank (in its earlier form IBRD) for capital aid, and the inter-agency fund known as the Expanded Programme for Technical Association (EPTA), set up in 1949 for the provision of technical experts and training. This programme provided funding for small scale development projects created by agencies of the UN system.

Principal among them of direct relevance for children were United Nations Educational, Scientific and Cultural Organisation (UNESCO) and World Health Organisation (WHO). The activities of other agencies were more or less relevant to children. For instance, the work of the International Refugee Organisation (IRO), latter to become the United Nations High Commission for Refugees, (UNHCR), had direct relevance for all children with refugee status.

The International Labour Organisation (ILO) continued to help children in various year. For example, they set minimum labour standards to discourage the exploitation of child labour or of nursing mothers. ILO established vocational training schools and systems in many countries and its technical assistance activities were also of value for children preparing for technical education.

The programmes of the Food and Agriculture Organisation (FAO) were directed at providing technical assistance for the development of agriculture, fisheries and forestry. While typical FAO projects are not directly aimed at children, the basic support for food production inherent in many of them could be seen to have a broad developmental impact helping developing societies and the children in them. In addition, their nutrition programmes are often for children and other malnourished groups.

UN world demographic statistics suggest that mortality is highest among infants up to five years, especially in less and least developed countries. The WHO Constitution, adopted in 1948, recognises this in stating that the 'healthy development of the child is of basic importance'. Since that time WHO proclaimed as its goals the attainment of health and of primary health care for all by the year 2000. WHO activities in the filed of nutrition, aid to nursing mothers and epidemic control, all have directly contributed to better conditions for children, especially in developing countries.

During the 1950s and early 1960s, the UN system evolved with the increasing application of capital aid from IBRD and its sister agency, created in 1961, the International Development Assistance (IDA). This capital was initially for infrastructure projects—bridges, roads, dams, electrical systems— all supporting economic development.

In the last ten years, however, the World Bank group has become involved in socially-oriented projects for least developed countries, including integrated rural and regional development projects, and technical educational support. Meanwhile, the UN system's technical and pre-investment assistance grew in dimensions and scale. To the initial work of the EPTA was added the UN Special Fund (set up in 1957). Both organisations were merged in 1966 to form the UN Development Programme, which continues to finance the largest part of technical cooperation activities mostly by specialised agencies. Many of these activities in various ways benefited children. Meanwhile, the UN system established new instrumentalities. The UN's own food aid programme, the World Food Programme, became increasingly and dynamically involved in the 1960s in assisting less privileged groups, including children.

In the early seventies the UN set up a fund for population activities, the United, Nations Fund for Population Activities (UNFPA) to assist countries requiring aid in all matters affecting population including demography and population stabilisation. The importance in the crowded modem world of limiting family size to foster the welfare of individual children needs hardly to be emphasised. Through its activities UNFPA has undoubtedly contributed to the welfare of children already born by helping in appropriate cases to limit family size. In an analogous way, the UN Environment Programme (UNEP) was set up at the UN Conference on the Human Environment in Stockholm during 1972. Its activities, aimed at promoting a better environment and limiting environmental destruction by hasty or ill-planned development, can result in an improved environment for children.

In recent years the UN system has worked increasingly to promote cross-fertilisation of experience among developing countries. The 1978 Buenos Aires UN Conference on Technical Cooperation among Developing Countries (TCDC) helped to bring out the need for and value of this approach. TCDC can yield benefits for countries promoting child welfare exchanges with other countries which have comparable experiences.

Role of UNICEF

Foremost, but not alone, among UN agencies and bodies concerned with helping children, UNICEF has played an important role in developing a 'country' approach to the child. By following the country approach, UNICEF has been better able to assist developing countries to formulate national policies for children. It has also cooperated closely with other international agencies such as UNDP, WHO, UNESCO, FAO and ILO, in helping channel more resources into meeting the needs of children and youth.

UNICEF encourages developing economies to build up national policies of child development, on the one hand, and induces the developed world and international agencies to increase aid allocated to child development programmes on the other. Generally, UNICEF has seen its own role as a catrlytic agent, giving peculiar attention to the instigation of programmes which should be capable later on attracting assistance from external, or development from internal, sources. Therefore, the inclusion of projects for which it is providing assistance in the national development plan (that is, where a recipient country has such a plan) follows as a matter of course. This is a method of ensuring that sufficient local support is being given to the schemes in which UNICEF may be participating. Once the decision was taken that the Fund would provide assistance on a country basis, a structure for financial and technical commitments oh a long-term basis was evolved, permitting projects of a long-term nature.

UNICEF has realised that, if assistance is to be geared to the national plan to a country, it will have to support combined community services. Of special importance is the training of workers of staff children's services. Moreover, except for limited opportunities for direct distribution, it is impossible to provide for children without providing direct and indirect benefits to their families. It has to be understood that the welfare of the young depends, primarily on the welfare of its mother, and throughout the welfare of children is closely interwoven with that of their families. UNICEF assistance programmes are based on an analysis of the needs of children in the country concerned within the context of the family and the community and hence are formulated on a cross-sectoral integrated pattern.

Thus, UNICEF has adopted an integrated, cross-sectoral approach in determining its activities. This has required it to call for technical assistance and cooperation from a number of agencies, leading, in turn, to a mobilisation of an increased volume of resources for child development in Asia, Africa and Latin America. During the second Development Decade of the 1970s, UNICEF continued to press for comprehensive planning to meet the needs of children and youth undertaken within the framework of national planning.

Direct Action for Children in the 1980s

Assistance for efforts to meet basic needs, nutrition services and education development is likely to continue during the International Year of the Child and through the decade of the 1980s for those countries requiring help. In addition, it is proposed that consideration be given to assisting developing countries in the following ways:

1. In formulating National Child Development Plans as an integral part of the National Development Plan and woven into, as appropriate, sectoral plans for health, education and welfare. This might be achieved by establishing a government planning post of adviser for child planning in the Office of the President or the Minister of Planning. It would be necessary to have direct links to the chief sectoral ministries concerned in Social Welfare, Education, Population and Health. On government request such posts could be financed from UNDP.
2. Within National Health Ministries establishing a post of Child Health Planner. Such a post could be concerned with public health planning or disease prevention at the school and village level, in conjunction with the Ministry of Education and local government organisation. Again, UNDP could finance such a post if a government requested it to do so.
3. Assigning responsibility for public information work to disseminate information on child development during the year and to plan specific 'events' to promote public understanding of this.
4. Special work is needed to reach the disabled child. In 1981 the UN will observe the Year of the Disabled. Between the Year of the Child and 1981, action programmes can be developed to help disabled children and to spread the experience from successful efforts to other countries.
5. Action should be taken by Governments, Information Ministries, Non-Governmental Organisations and media organisers, to ensure full coverage for the Year of the Child and highlighting of the issues of development as they affect children—poverty, bad

housing, poor water supply, inadequate environment. A poor environment is a function of poverty. Development efforts aimed at improving the urban environment through slum clearance, better city planning and construction, will surely help advance the objectives of the International Year of the Child.

6. Much is known by now of the success of town twinning and the extent to which cities, in Europe especially, have developed links with other countries. In like manner, it is proposed that schools establish a 'twinning' relationship. This can be done and developed through personal contacts; those who lack contacts may wish to write to the Ministry of Education in the country where such an association is sought.

In the case of developed countries, it is urged that schools try to develop at least one 'twin relationship' with a school in a developing country. Such links have been developed already by schools in England (such as Ampleforth College with schools in other parts of Europe), and by various schools in North America. In the public schools of West Haven (Connecticut) steady links with schools in Australia, Colombia, and the Philippines are nurtured and personalised by the exchange of students with those countries for a whole school year.

The Future

'The past is prologue'. Most of the above-mentioned activities of the UN system of organisations will necessarily continue as long as the basic problems which they address persist. A key problem will be that of resources. Resources for UN technical assistance and aid for children and for development generally, depend on voluntary contributions by Governments. As of October 1978, the UN received about $850 million in pledges for 1979, for the two major sources of funding of development activities involving children, i.e.

UNICEF and UNDP. This amount is separate from the resources available to the World Bank Group for capital loans. The total resources of the UN system for direct grant assistance to developing countries may thus (with optimistic projections of rates of growth in voluntary pledges or contributions) reach $2 billion annually by the beginning of the Third World Development Decade in 1981. Compared to the widely publicised $400 billion spent in 1977 on arms by ail countries, the amount for international cooperation against poverty and for development is tiny, even if bilateral aid (totalling around $45 billion in 1977, according to OECD) is added to UN direct aid. Even these totals looks thin once averaged out and considered at the national level of pressing needs in more than 130 developing countries. A recent report by the UN Secretary General warned that 400 million children might suffer brain damage through malnutrition if rapid action were not taken to avert near famine conditions. Many children in the Third and Fourth World are afflicted with poor diets, endemic disease, bad environment, inadequate or non-existent education. The latent disharmony in resources endowment, relative levels of prosperity, and access to growth opportunities between the North and South of the earth, means that the United Nations as an organisation faces an awesome choice.

REFERENCE

UN ECOSOC E/ICEF/627, *Children in the Second Development Decade: Priorities for Planning,* New York, April 1973.

7

Rights of the Child and NGOs

> "A mother once asked a clergyman when she should begin the education of her child, who she told him was then four years old. "Madam," was the reply, "you have lost three years already. From the very first smile that gleams over an infant's cheek, your opportunity begins."
>
> —*Abigail Whately*

It is pertinent to note that Government of India in 1974 adopted a National Policy to provide proper protection and services to children to ensure their mental, physical and social development. Later on the UN Convention on Rights of the Child was adopted by the General Assembly of the United Nations on 20th November 1989. By the end of 1993, 189 countries had signed the Convention. The UN Convention of the rights of the child is unique. It advocates measures for the protection, natural development of the children, their survival and participation. But that is an ideal which only a few nations have achieved so far.

In spite of the fact that in India the Government had realised the pitiable plight of the child much earlier and had enacted laws for the protection and welfare of the child, it is rather painful to know that more than seventeen million children of independent India under the age of 14 are working in most inhuman conditions. No proper food, no sanitation, no medical aid and no education. The laws are there, but they are not implemented.

These children do not know what childhood is. They do not even know what games in the open air one could enjoy playing. My personal observation of children working in the carpet industry, bangle making factories and bidi factories makes me feel most depressed. These children are working for long hours on very low wages. The parents are obliged to send them to the employers to work under these inhuman conditions for the survival of the family, as the income of the head of the family is too low to meet the basic requirements of the family.

Ironically, on one hand we feel concerned about the fast growing unemployment, on the other hand we put the children to work which creates unemployment for the grown ups. If the father-mother get the minimum wages fixed by the Government, they can look after the needs of the family and the children would not have to work. Child labour must be discouraged. Statutory minimum wages fixed for men-women must be paid and here the NGOs can play the role of watchdogs. Not only this, they can also bring about awareness among people, motivate the parents not to allow their children to work, rather put them in school.

They must be made literate, and trained to be able to look after their needs later in life. With governmental support NGOs can approach industries to prevent employment of children under 14. They can also help to ensure that adults get their minimum stipulated wages. Where workers have to live in industrial complexes, the children's schooling should be made compulsory in the factory area.

Facilities for primary education have grown more on paper and in name than in reality. In the villages where I have been visiting with my colleagues we often find that either the teacher or the peon is absent and sometimes both. Children come, and finding the school closed, loiter about and go back home. At the end of the year there is bargaining between the parents of the child and the school. The child gets promoted in exchange for money. This results in total illiteracy inspite of the existence of primary schools.

Here too the NGOs can play the watchdog's role, report to the concerned departments and improve the system. The problem is that the authorities concerned show little interest in taking action. Interaction between NGOs and the concerned department is a prerequisite. It has to be cordial and constructive. Rural children hardly get any medical help. In remote areas of our rural India we can do something to help and save the life of many children who die for want of medical aid. If there is some rapport between NGOs and the health department, the welfare organisation can do much to save many young lives. We all know that many children die before the age of six and before that many die during the teething period. Proper nutrition and proper medicine is the right of every child. Here too the NGOs can play an important part by creating awareness among the parents and also by insisting that the father and mother must not be deprived of minimum fixed wages, which can enable them to save the situation.

In fact, there are many other areas of child abuse. Children are used by boot leggers and drug peddlers. An innocent soul is put on the most hideous type of work. Children are trained for pickpocketing. They are made use of in thefts. These practices have been going on unchecked. Hardly any police help has been available. NGOs can definitely play an important role with the help of the police to check such evils. Many unfortunate children are abducted and made to beg by gangs which have made this into a money making industry.

In some cases the child's limb or eye is purposely damaged to make him look helpless and pitiable so that he can fetch more money from begging. Gang hideouts can be detected by NGOs with the help of the police. In female prisons some time we have seen young children staying with mothers under the most undesirable atmosphere and under inhuman conditions. In Calcutta, Bombay and Lucknow our convenors have succeeded in getting the children released from jails and they are then rehabilitated. The NGOs can help the innocent child by getting him out of the criminal surroundings of the jail where he is kept with the mother. Here too we need the sympathetic cooperation from the jail authorities.

In short everywhere, in every problem of the child, NGOs can do some constructive work. But for this, interaction with the Government and the concerned authorities is a necessary precondition. Because of its helplessness and vulnerability the child is the worst sufferer in spite of all the "National Plans for the Health, Happiness and Survival of the Child". Let us all: the concerned Government departments, the police and the NGOs and Welfare Organisations join together to protect the Child and its Rights. The children of a country are its most valuable asset. No nation on this globe can ignore the responsibility to ensure the proper growth and development of children as the future of the country lies with them. However, as a voiceless section of the society, the children have suffered throughout the ages at the hands of those who are supposed to provide them with security and love. It is not uncommon to find children who have been subjected to abuse, torture or exploitation and who are denied their basic rights. Children are the worst sufferers of wars, natural disasters, unemployment, poverty etc. It would be worthwhile here to cite few startling facts.

According to the Human Development Report (1994), in developing countries, more than two million children die annually of preventable infectious diseases and 192 million are malnourished. The report also reveals the Human Development Cost of Arms imports. According to the report, India ordered 20 MIG-29 fighter aircraft from Russia at a cost that could have provided basic education to all the 15 million girls out of school. Thus, increased expenditure on defence results in denial of the basic rights to the defenceless section of the society, i.e. children. The fact that the children are physically and mentally immature necessitates special rights for their nurture, protection and fulfilment of unique needs. The rights of the child should be a matter of great concern to one and all as the children constitute one of the most vulnerable sections of the society.

A deprived child may belong to any religious, ethnic, linguistic, socio-economic category and hence universal acceptance and promotion of child rights needs greater emphasis than the rights of other deprived sections of the society. The Constitution of India under the directive principles of State Policy has laid stress on formulation of policy by the State towards prevention of abuse of children, provision of opportunities and facilities for healthy development and protection of children against exploitation and moral and material abandonment.

The Indian Government has also in pursuance of its policy, made continuous efforts in the direction of ensuring fair and just rights to children through enactment of legislation, framing appropriate policies and providing special services to the children. In spite of these initiatives, there is little achievement to our credit. According to 1981 census, the number of orphan children below 14 years of age was 37.59 million. The Government of India estimates of working children show that there are around 18 million working children in India whereas a study conducted by ORG estimated the number of child workers at 44 million.

About 90 per cent of working children live in rural areas. Thus India has yet to do a lot to ensure prevention of deprivation. The ratification of UN convention will go a long way in protection and promotion of just and humane rights of children.

Role of NGOs

It is significant to note that non-government organisations (NGOs), working with children have been the great source of relief especially to the deprived children who live in exceptionally difficult circumstances. NGOs have made untiring efforts to provide for the basic necessities of life to hitherto most vulnerable group not only through custodial care and protection but also through their developmental approach and participation in policy making. Looking at the magnitude of the problem, NGOs, however sincere and committed may their efforts be, have not yet been able to reach all the needy children.

The reasons are manifold: scarce resources, inadequate trained manpower, lack of cooperation from government and other bodies etc. However, in the light of the UN Convention on the Rights of Child, NGOs do have to play greater role at different levels. In the present scenario, this would require redefining their roles, change in approach and global thinking. Acknowledging the fact that NGOs, especially working in the field of human rights, have excercised their influence on the content of convention, the implementation of the same can hardly be effective in absence of participation of voluntary sector in the observance of the provisions of the UN Convention.

The NGOs have primarily concentrated on provision of basic amenities to the deprived children while functioning at the grassroot level. Most of the NGOs have made available to the needy children, the services which are better in terms of quality as well as effectiveness. Several NGOs have made effective use of the available resources and with most appropriate methods generated resources on their own. However, besides service delivery, there is also a need for education of the community, particularly at the grassroot level regarding its responsibilities towards children in need of help.

Not many NGOs have taken up education of people a grassroot level in an intensive manner. As a result, care and protection of children living in exceptionally difficult circumstances is perceived as the sole responsibility of the

NGOs whereas the primary responsibility lies with the family and the community. There seems to be overdependence on NGOs for identification, care, protection and development of deprived children. The NGOs in turn find it difficult to cope up with the ever increasing clientele given the inadequate resources at their disposal. Awareness at grassroot level will bring about increased sensitivity on the part of the society to understand the problems of deprived child and would also help NGOs share its responsibilities with community.

Here, the NGOs can strive towards achieving what is popularly known as "synergy" through better co-ordination of services at local and regional level. We often come across examples where more than one NGO functions in the same geographical area, with same clientele and provide more or less similiar services. NGOs working for children are no exception to this. A better approach would be to cater to the different clientele or providing different services to the same clientele in a geographical area.

This would require local networking through promotion of joint forums. A remarkable example in this regard is the Forum for Disability formed by NGOs working with disabled in Baroda. In respect of NGOs working with children or having children as one of the clientele, such forum will avoid problems of overlapping of its services. Apart from this, NGOs will have to be prepared for increased participation in regional/local planning bodies as well as policy formulation with regard to the children.

The NGOs, besides being supportive, can also take up advocacy and lobbying for children's rights, particularly at such places where there are increased violations of such rights. Not many NGOs have taken up this as one of the key functional areas. If NGOs are able to pose a united front, they would emerge as a potent force for several initiatives in the area of child welfare and development

Need for Global Thinking

Remarkably, "Think Globally and Act Locally" has been a popular slogan till now. The problems of children cut across the boundaries of region, nation, caste, creed as well as colour. Although the magnitude as well as nature of the problems may vary from region to region and from country to country, there exists some commonality either in philosophy or future vision. By the end of 1993, 159 countries had either signed the Convention or had become state parties to it by ratification.

The NGOs of these countries can interact with greater frequency through international bodies like UNICEF and share their experiences, collaborate with each other on the areas of common concern, transfer the expertise available with each other and thus make the world better place to live for children. The NGOs at present, no doubt, interact internationally at the initiative of International agencies, frequency of such interaction with emphasis of collaborative approach can increase.

The specific areas where NGOs have to intensify efforts also need some attention. The UN Convention on the Rights of the Child draws attention to four sets of civil, political, social, economic and cultural rights of every child. These are in the form of the right to survival, the right to protection, the right to development and the right to participation. The priority areas for NGOs in respect of each of these are discussed below.

The first and foremost issue that NGOs have to address is disparity in boys and girls ratio. According to "the Right to be a Child", a UNICEF India background Paper, there are neatly 4 million "missing girls" in the age-group of 0 to 6 years which is a clear indication of female infanticide—a cause for such adverse sex ratio. Alongwith this, the problem of avoidable child deaths i.e., high infant mortality rate also deserves immediate attention.

The present rate of Infant Mortality at 79 and the under-five mortality rate of 124 per 1000 births shows a high prevalence of child deaths due to preventable and environmental causes. It is an established fact that parental education particularly of women and availability as well as utilisation of health services would bring about remarkable changes in IMR and increase chances of children surviving to a great extent. NGOs can focus attention on the above two areas. Better provision of health-care services can be ensured through active participation of NGOs in health-care delivery system, particularly preventive health-care, especially in the areas where child survival rate is low.

Although provision of primary education has been the responsibility of the state, Article 45 of the Directive Principles of State policy as well as the National Policy on Education reflect the firm commitment on the part of state to universalisation of free primary education. There is substantial evidence stating progress in the provisioning of primary education in terms of increase in number of primary schools, school teachers, enrolment but there is paucity of reliable data on quality of education and attendance. The role of NGOs in this regard would undoubtedly centre around providing supportive services as a result of which *(a)* Girls can avail of the same opportunities as are available to boys *(b)* The dropout rate is reduced particularly among the disadvantaged sections of the society *(c)* assistance to government machinery in better allocation of resources, and *(d)* help in making education a meaningful experience to the children.

The right to protection implies prevention of exploitation, abuse and torture which may endanger child's education, health, development or survival. India has a long history of economic exploitation of children who are employed as labourers. Poverty and lack of legislative measures to make primary education compulsory drive the children out of school to work in conditions most unfavourable to their growth and development

The role of NGOs would therefore be to create awareness among masses against the evils of child labour. Moreover, such awareness should be aimed at building public opinion for elimination of child labour. The crucial task for NGOs therefore would be lobbying and advocacy against exploitation and abuse of children in varied forms.

The right to participation includes the right of accused children to be treated in just and Humane manner. Although very limited information on juvenile crime and the treatment meted out to such children is available, there is urgency of action in respect of inevitability of adequate resources to provide for legal representation of accused juveniles, facilities for separate custodial care, rehabilitation services etc. The issue of child marriage is also on forefront which demands intervention by NGOs.

The NGOs can work on the core areas by playing the role of educator as well as activist. In order to ensure fair treatment to the children under custodial care, the NGOs may bridge the gap that exists between general public and children under custodial care which would make worthwhile difference not only in terms of nature of care provided but also safeguard interests of the children. With regard to child marriage, NGOs working in the areas where prevalence rate is higher, can take up the issues through public participation in eradication such evil, seeking cooperation of judicial and executive machinery.

Thus, NGOs have vital role to play with regard to basic and just rights of children. No country can fulfil the rights of children without active participation and cooperation from NGOs. The need therefore is to delineate the scope of NGO's functioning and making coordinated and concerted efforts towards ensuring children a bright future.

The UN Convention on the Rights of the Child is the biggest landmark in the sphere of human development at the national and international level. The Convention came into force on 2nd September 1990 and was ratified by India on 12th November 1992.

The Convention had been endorsed by the World Summit on Children. In fact this was the first major global action for the implementation of the Convention. In pursuance of the commitment made at the World Summit, the National Plan of Action for Children thus prepared provides the framework for actualisation of the Convention.

The Government of India is making sincere efforts to meet the provisions stated in the Articles of the Convention to give to the children their rights. Article 42 of the Convention states that State Parties shall undertake to make the principles and provisions of the Convention widely known, by appropriate and active means, to adults and children alike.

NGOs have been a partner to the Government in implementing programmes related to welfare and development of children. It was suggested by the Department of Women and Child Development that NGOs may be actively involved in implementing the Convention. It was felt that for creating awareness and getting inputs from NGOs for this endeavour, workshops may be organised at the state and national levels.

With this in view, UNICEF and the Department of Women and Child Development organised 11 state level workshops in collaboration with ICCW and its state branches and selected NGOs. The objectives of these workshops were to:

(i) orient NGOs on the significance of the Convention on the Rights of the Child;

(ii) identify problems and issues relating to situation of children in general and the girl child in particular;

(iii) analyse problems and suggest solutions for implementing the Convention on the Rights of the Child; and

(iv) identify areas for action which could be undertaken by NGOs.

During these workshops, very useful recommendations were made regarding the implementation of the Convention. In the present document, Role of NGOs in Implementation of the Convention on the Rights of the Child: Workshop Recommendations, efforts have been made to collate all the recommendations made in the workshops held in Andhra Pradesh, Bihar, Chandigarh, Karnataka, Madhya Pradesh, Maharashtra, Orissa, Rajasthan, Uttar Pradesh and West Bengal. The list of workshops held appears at the end of the recommendations.

The recommendations are grouped under specific areas of action i.e. advocacy, child abuse, child care services, child labour, children in difficult circumstances, community participation, co-ordination and networking, culture, disabled children, disasters and children, education, employment, family, girl child, health, institutional and non-institutional services, juvenile justice system, laws and legislation, monitoring, evaluation and research, nutrition, nutrition and health education, participation by children, plan for implementing rights of the child, policy and planning, pre-school education, recreation, role of media and role of NGOs.

As NGOs have an active role to play in the implementation of the Convention, this document can serve as the basis to initiate action to promote programmes to ensure children the rights related to survival, protection and development and participation.

Recommendations

1. Advocacy

NGOs should undertake advocacy and lobbying to defend the rights of the child. (ORI P6)

> A Plan of Action for advocacy and dissemination of information regarding the Convention on the Rights of the Child (CRC) should be formulated. (WB P3)
>
> Advocacy efforts to create an action oriented opinion on the CRC is to be operationalised. (WB P5)

NGOs should form pressure groups to motivate the Government to implement CRC. (KAR P4)

Greater attention needs to be paid to advocacy. This should be targeted at policy makers, trade unions, public, legislators, etc. (MAH P8)

2. Child Abuse

Regarding the issues of physical, sexual and psychological exploitation, the community should be motivated to intervene in cases of such abuse alongwith those meted out to girl children. Community involvement in identification of cases like child marriage, dowry-related abuse and so on should be planned. (WB P4)

Sex education in schools should be initiated and discussions on sexuality and protection from sexual abuse should be encouraged in schools. (WB P4)

NGOs should encourage investigative reports on incidents of child abuse and violation of child rights. (KAR P5)

It is necessary to eradicate various forms of child abuse like child prostitution, child labour, child beggary etc. with the help of laws. Arrangements for rehabilitation of child prostitutes is also very necessary. (UP P46)

All possible measures should be taken for prevention of abuse and exploitation of the institutionalised child by opening up visibility and encouraging interaction of communities and institutions. (AP P24)

Enforcement authorities, NGOs and community at large should be adequately empowered to take all measures to protect children from all forms of physical, mental and sexual abuse and exploitation. (AP P24) (MAH P8)

3. Child Care Services

The working hours in anganwadi centres and day care centres may be extended to take care to children after 1 p.m. where mothers go for employment in rural areas. (AP P16)

More creches should be opened at work centres of adults. Illiterate parents should be discouraged from giving sedatives like opium, etc. to infants. (AP P21)

4. Child Labour

A State Level Commission should be set up to work for eradication of child labour. (CHA P8)

There should be an Apex Committee to campaign against child labour at regional as well as State level. (CHA P8)

Children of organised women bidi workers get scholarships/uniform and medical facilities, whereas children of unorganised workers are deprived of such facilities. Therefore, NGOs can organise parents of child labour in their respective regions and make available to them adequate facilities and benefits of various schemes such as Jawahar Rozgar Yojana, Small Scale Industries etc. (MP P9)

For dealing with child labour the relevant Act should be modified to cover all child workers and not only those employed in the organised sector with the ultimate objective of banning child labour. There should be a ban on the use of child labour in households. (WB P4) (AP

Primary education should be made compulsory and the children should be in schools instead of in work places and whatever measures are required must be guaranteed. (WB P4)

Child labour should be banned as soon as possible

and families where children are the sole earners should be provided alternate means of income. (UP P47)

5. *Children in Difficult Circumstances*

In the case of a child liberated from difficult circumstances, proper care has to be bestowed on creating alternative schemes and support systems for the child to see that he/she does not go back to the same or end up in a worse situation. (KAR P4)

Adoption services, foster homes, residential institutions etc. may be started for children born outside wedlock, abandoned children and children who have lost contact with parents in natural calamities, disasters etc. (AP P15)

6. *Community Participation*

To achieve the goals set forth in the Convention on the Rights of the Child, Mandals should be formed involving

— Children

— Youth : Boys and Girls (Yuva and Yuvati Samoohs)

— Women (Manila Samoohs)

(RAJ P31)

Community organisations should be formed to ensure the enforcement of rights related to child protection. They should be given identity cards and necessarily authorised to work for protection of child rights (UP P46)

7. *Co-ordination and Networking*

There should be a common platform for addressing common issues and guiding direction about work. (CHA P8)

There should be close coordination between various Government Departments and NGOs and accountability to each other with concurrent monitoring and evaluation to achieve the Mid Decade (1995) and the Decadal Goals (2000) already set by the State and Central Government. NGOs may play a more active role in monitoring and reporting. (BIH P40) (AP P23) (MAH P8)

Donor agencies should fund NGOs directly for small projects not exceeding Rs. five lakhs and undertake monitoring and evaluation directly. (BIH P30)

There should be direct relationship between NGOs and UNICEF for programme implementation (BIH P45)

Coordination between all agencies working in related areas of rights of children is essential, including groups/NGOs working with women and Dalits. (MP P11) (KAR PS)

Different Government committees working for children should have NGO representation. (WB P4)

There should be a Suggestion Cell where NGOs can put forth their problems and suggestions (WB P4)

Links should be established with prominent members of the village for example Gram Sevaks, ANM, AWW, Teachers, Dais, Ranch, Sarpanch, influential men and women of the villages and NGOs in the villages. Links should also be established at the block level e.g. with BDO, Government Officials at district and state levels. (RAJ P31, 32) (KAR P5)

A Coordination Committee organised by NGOs should be formed at the district level. In this Committee all NGOs with good contacts should be involved. (RAJ P32)

A Resource Team should be formed so that training can be organised in the villages for village workers, the coordinating group and the pressure group.

This team should identify methods, agencies and government organisations. Information may also be given on how to start dialogues with children, their understanding of things etc. (RAJ P32)

8. Culture

The syllabus in schools should include lessons on the composite culture of the country. (AP P20)

Teachers should be trained and made culture conscious at the state or central level. (AP P20)

9. Disabled Children

Special attention should be given to handicapped children and children belonging to special categories. (UP P45)

More hostels need to be set up for disabled children in proportion to their numbers. (AP P17)

Specific budget component should be alloted for disabled children. (AP P17)

10. Disasters and Children

There should be a special department to deal with the protection of children in special circumstances like manmade disasters such as riots, terrorist activities, wars etc. and natural calamities like floods, earthquakes, famines etc. and these functionaries should be trained to deal with all aspects of such situations. (UP P46)

11. Education

Primary education should be made compulsory by law and available tree to all. (WB P4) (AP P13, 22) (UP P44)

Free and compulsory school education for all children upto the age of 18 years in pursuance of the CRC is the need of the hour. Relevant quality education for all children, including the physically and socially handicapped must be ensured. (AP P24) (MAH P8)

The Government may provide some incentives to school going children such as hostels, free food, clothes and books. Families in need may be provided monetary assistance/scholarships for their children. (AP P13) (UP P44).

To ensure cent per cent enrolment and retention, the Mid-day Meal Scheme should be implemented in rural, tribal and urban slum areas. (AP P19)

Timely and adequate supply of educational materials should be made to children living in hostels so that they do not face hardships or punishment for not doing their work. (AP P17)

Quality of education should be improved by providing adequate training to all teachers. Regular and periodic orientation courses should be organised for the teachers especially in the rural areas (BIH P25) (AP P19)

Special training should be given to teachers of private schools and recognition accorded to neighbourhood schools so that children continue to get education in the formal system. (BIH P25)

Concentrated efforts need to be made in hilly and tribal areas to reach education to the unreached. (BIH P25)

There is need to develop a cadre of voluntary teachers and barefoot teachers to reach remote and inaccessible pockets. NGOs can streamline such activities by paying a small honorarium to such teachers who carry education to the doorsteps of the unreached. (BIH P25)

Curriculum should be redesigned and learning environment made enjoyable to make education attractive to children, parents and teachers. The syllabus should be more relevant and must prepare the children for real life situations. (BIH P25) (AP P13, 18) (UP P45)

Education should not be linked only with employment. It should be accepted as a means for overall development of the child and for improving the quality of life. (BIH P25)

At present education is considered as Government's programme and responsibility. The focus should be reoriented making education as people's programme. (BIH P25)

Educational arrangements in villages and towns should be such that schools are situated at a walking distance for children. (MP P9)

Schools should also have creches as many children stay back at home to look after their younger siblings. (MP P9)

Adequate teaching staff and teaching facilities should be provided. In addition, facilities like toilet, clean potable water and playground should be made available in schools. (MP P9)

School curricula should include provisions of laws relating to the protection of children's rights. (MP P11) (MAH P8)

Measures must be taken to ensure that school discipline is administered in a manner consistent with the child's human dignity and conformity with the present convention. (AP P13)

Education should develop the physical, spiritual, mental and behavioural personality of the child and his inherent talents. (UP P4S)

At the State level, the budget allocation for expenditure on education should be increased. It needs to be given top priority to achieve Education for All by 2000 A.D. Expansion of social welfare hostels to accommodate motivated children is needed. (UP P4S) (AP P24)

12. Employment

After a minimum of 14 years, children can be trained in life skills. In this area, Government can replicate some successful NGO models. (KAR P4)

Fifteen to 18-year-old children should be provided vocational training so that they can earn their livelihood in future by self-employment or in some vocation. The trades taught in Vocational and Guidance Centres should be relevant to the needs, aspirations and requirements of the local area. (UP P45) (AP P15, 19)

13. Family

In any programme concerning children's welfare or empowerment, the child should never be looked at in isolation, but as an integral part of the family. (KAR P4)

In the school syllabus, the role of parents and other members of the family in the changing social set up may be depicted appropriately at a suitable level. (AP P15)

For children of prostitutes and children whose parents are undergoing imprisonment or chronic patients, efforts should be made to re-establish family ties so that at a later stage the child can have an identity and support of the family. (AP P15)

Children from battered and violent families are likely to have disturbed and stunted growth and acquire deviant behaviour patterns, hence family counselling centres may be increased in number and run by Government or NGOs with qualified and skilled counsellors so that couples can seek guidance in case of disharmony in the family situation. (AP P15)

To ensure a congenial family environment for the healthy growth and development of children, pre-marital counselling centres need to be opened. (AP P15)

14. Girl Child

An integrated approach for the empowerment of girl child should be adopted through a single window delivery system and not through segregated interventions for health, education and income generation. (BIH P45)

Coordination committees at every important administrative level like Block, District and Division of the Government and NGOs should be formed for effective planning and implementation of the girl child's welfare programmes. (BIH P45)

Sports and karate needs to be encouraged especially for girls, the latter may also help in self-defence. (AP P17)

More facilities should be provided for education of girls. (UP P45)

Girl children in 14-18 years age group should be given vocational education and value oriented education. (BIH P45)

Appropriate legislation should be adopted for effective protection of the girl child. (BIH P45)

Sex related discrimination should be prevented and the practice of sex determination tests should he banned. (WB P4) (UP P44)

15. Health

There is a need for improving referral services. (CHA P8)

Success programmes, such as Medical Melas, i.e., catch up round for immunisation and for comprehensive health care, should be implemented in all districts using district level NGOs. (BIH P22)

Preventive and promotive aspects of health services should involve NGOs in a big way and effective manner by identifying some nodal NGOs in every region of Bihar. (BIH P45)

Well operationalised surveys on immunisation status of children should be conducted by UNICEF through various NGOs all over West Bengal. (WB P4)

Health department and NGO cooperation should be introduced in all programmes like MCH, family welfare and primary health care in rural and urban areas. Mobile Health Care Units too should be introduced. (WB P4)

It should be ensured that health services are available to people, specially the weaker sections, at their doorstep free of cost through improved Public Health Delivery System. (UP P44) (MAH P8)

Mobile vans should be provided to conduct health check up at every school atleast once a month. (AP P17)

Measures should be introduced to regulate birth rate. (WB P4)

16. Institutional and Non-institutional Services

Infrastructurel support in the form of Homes and better living conditions within these Homes should be ensured. For such infrastructural support there should be the provision of a Child Budget. (WB P5)

Common procedure may be evolved for all children's institutions run by any Government Department or voluntary organisation whereby food is provided according to scales determined in consultation with the National Institute of Nutrition and other requirements like plates, mugs, clothing, bedding, soap, etc. are issued according to needs without any monetary limit. Sufficiently adequate budget allotment to meet the minimum needs of children, i.e. food, shelter, clothing, education, water and electricity should be allotted to institutions. (AP P14, 16)

Non-institutional approach for child rehabilitation should be given priority over institutionalisation of the child. (AP P24)

Positive cases of child care in institutions should be highlighted. (AP P24)

17. Juvenile Justice System

Officials of the Judiciary and Police are to be oriented and updated on the Articles and obligations of the Convention on the Rights of the Child. (B1H P28) (UP P46)

Sustained efforts should be made to disseminate the importance of CRC with the general public to mobilise community support (BIH P28)

Anti-social elements, gangs and groups who induce children to commit crimes should be brought under law and punished. (BIH P28)

Juvenile Courts should be established in each district under Juvenile Justice Act 1986 and a full time judge should be appointed. (MP P11)

Children can avoid the traumatic experience of going through the court procedure if the stipulated number of Juvenile Courts, Juvenile Welfare Board and Family Courts as provided in the Juvenile Justice Act 1986 are set up. (AP P15)

Orientation training for all persons engaged in the

field of rights of children should be organised compulsorily. (MP P11) (UP P46) (MAH P8)

There is a need to set up a Monitoring Cell which will coordinate the activities of the Legislative bodies, the Judiciary and Police. This Cell would also hold these different sectors accountable for any breach of duty. (WB P3)

Together with raising awareness on the CRC among the law makers and enforcers, it was also necessary to disseminate information to the public alongwith orientation and training of judiciary and police to deal with child offenders in a better manner. The Juvenile Justice System needs to be reviewed and revised to make it child friendly. (WB P3) (MAH P6, 8)

For children in conflict with the law there should be efforts to solve minor offences at the community level without going to the Courts for trial. (WB P4)

Juvenile Courts should take the help of NGOs in investigating and deciding cases and the police too should be sensitised and made aware of the techniques of dealing with children. (WB P4)

There is a need for greater awareness on the provisions of the existing laws and their effective implementation (WB P5)

The composition of the Juvenile Board should be carefully planned so that both the members and magistrates have special knowledge and orientation to child psychology. (WB P5)

Compulsory Legal Aid to child offenders should be provided and a separate panel of advocates supported by social workers to assist and defend children should be constituted. (WB P5)

Mobile Juvenile Courts should be instituted to decrease the delay in the trial process. (WB P5)

A special trained police force with an investigation agency should be provided so that the police are capable of responding sensitively to child offenders. (WB P5)

A Monitoring Committee should be formed at district level with Judiciary, Police, Executives and NGOs for monitoring ongoing cases in court. (WB P5)

More Child Guidance Bureaus may be started by Government Departments and NGOs in rural and urban areas as the provide healthy alternatives to family care. They should also undertake the responsibility of education, training and rehabilitating vagrants, beggars and street children. (AP P14)

18. Laws and Legislation

Definition of age of child varies from 12 to 18 years under different Acts relating to children. The anomaly of varying age in defining a child needs to be corrected, uniform definition should be adopted and one comprehensive state legislation on issue of child labour needs to be enacted. (AP P13, 23)

Close monitoring may be undertaken by NGOs regarding the implementation of laws for protection of child rights. (BIH P28)

Like consumer forum, legal cells should be formed to protect children from cruelty and abuse. (BIH P28)

NGOs should function as Pressure Groups for effective implementation of laws protecting children. (BIH P28)

For the protection of rights of children, a state level committee should be constituted which consists of Judges, Police, NGOs, Government and other related people. (MP P11)

A single Court should be set up for disposing of all the cases concerning children registered under various acts. (MP P11)

Arrangements should be made for effective enforcement of laws relating to child prostitution, adoption and child labour. (MP P11)

Cases relating to children and especially those of child molestation or rape may be tried in camera and decided within a stipulated period of six months. (AP P15)

An act should be promulgated to take drastic action against sexual child abuse. (AP P22)

Arrangement should be made to ensure within stipulated time the implementation of rules framed by the Central and State Governments for the benefit of children. (MP P11)

Laws pertaining to children should be more effective and humane and a procedure introduced for accountability on the part of the law enforcers. (WB P3)

Laws and legislation have to be either amended or strengthened to suit CRC. Existing legislation vis-a-vis children needs to be reviewed and revised to ensure that all the articles in the CRC are justiciable. (KAR P4) (UP P46) (MAH P8)

The need is not to formulate more laws but to ensure better implementation of existing laws with superior infrastructural support and greater awareness. (WB P5) (UP P46)

Awareness of parents, employers, school children, inspectors and the general public need to be increased on the CRC and child related laws. Information about laws related to children should be clear at the national, state, block and village levels. (WB P4) (UP P47)

A National Law may be planned for protection of children based on the articles of the Rights of the Child. A comprehensive local law should be enacted as International Law will not be taken, into cognisance by Indian Courts and present laws relating to children are difficult to coordinate sectorally. (WB P4) (AP P23)

19. Monitoring, Evaluation and Research

For effective evaluation and monitoring of the programmes 'Progress Indicators' may be developed which promote unbiased results. For example, the Infant Mortality Rate, School Retention and Child Labour Status could be the indicators for the Rights to Survival, Development and Protection respectively. (WB PS)

NGOs should monitor the Government's programmes for effective implementation of CRC. (KAR P4)

Studies pertaining to the status of health, education, child labour, children in difficult circumstances should be undertaken to collect the baseline data to identify focus areas and to work for the implementation of CRC on a needbased strategy. (BIH P19)

Studies related to amniocentesis should be undertaken to understand the real situation of the girl child in Bihar. (BIH P19)

We do not have realistic baseline data on subjects like child labour, child abuse and drawbacks in the legal system particularly in juvenile justice. Studies should be undertaken to collect data in specific areas identifying problems and target areas. (BIH P28)

A self monitoring system should be developed. (RAJ P34)

20. Nutrition

It should be ensured that nutritious food is available cheaply and easily from various centres providing nutrition services keeping in view its importance and need during different seasons. (UP P44)

Greater emphasis should be given to the effective implementation of nutrition schemes for children, especially in tribal areas. (MAH P8)

21. Nutrition and Health Education

NGOs have individual image and credibility with the community. Right type of NGOs should be identified and their services utilised effectively for NHE by giving proper training, guidance and continued support. (BIH P22)

Health education, including water and sanitation, should be included in the school curriculum and a systematic School Health Education programme should be implemented using the services *of* NGOs (BIH P22)

People should be provided extensive education on nutrition and health education in local language. This work could be undertaken by educational institutions, media, other centres/ institutions and organised industries/trades. (UP P44)

22. Participation by Children

State Parties shall assure the child the right to express his/her views freely in all matters affecting the child and ensure respect for the opinion of the child. Their views may be given due weight in accordance with the age and maturity of the child, especially for victims of natural disasters, girls for consent before marriage, etc. (AP P13)

Children's Development Councils comprising only children below the age of 18 years may be formed at district and state levels with the District Collector and the Commissioner of Women Welfare and Child Development as the Chairman and Convenor respectively. Official recognition should be accorded to such councils. These councils may serve as Advisory, Monitory and Supervisory bodies with regard to policy and programmes concerning children especially in institutions. (AP P24)

In issues concerning children and their rights, the government should seek children's active participation at all stages of planning, discussion and consultation. (KAR P4)

There should be orientation and training programmes specially for decision makers and the children themselves. (WB P5)

Children should be given the opportunity and provided with a platform to express their thoughts, feelings and opinions using plays, games, paintings, essays, etc. (WB P5) (MAH P7)

Children should form their own associations to get their rights. (KAR P5)

There should be child-to-child inputs on information regarding child rights. (KAR P5)

In matters concerning children, children should demand for their participation at all levels, from planning, discussion to decision-making. (KAR P5)

23. Plan for Implementing Rights of the Child

A Child Rights Commission at Central and State level should be constituted on the lines of Human Rights Commission with necessary machinery for strict implementation of CRC with investigative powers. (AP P24)

Separate Ministries for child development may be formed at State and Central levels so that the main focus will be on children. (AP P24)

A single department may handle all programmes for women and children instead of the present system of Social Welfare Department and Department of Women, Child and Handicapped Development. This department should also be the nodal department to handle CRC. (MAH P7)

Rights of children in institutions need to be protected as a special area. (AP P24)

The State Plan of Action for Children (SPAC) must be more widely discussed with NGOs all over the state. Workshops should be held at the district level in Marathi to ensure better outreach. (MAH P4)

Training and sensitisation of officials at all levels in the fields of law, police, labour, medicine, education and in relevant government departments on the CRC should be done. A chhd-sensitive approach should be used. (MAH P8)

Increased allocation of budget must be made in social sectors, i.e., Department of Women, Child and Handicapped Development, Education, Health, Social Welfare. (MAH P8)

The Government must strive to create more awareness among the public on the Convention on the Rights of the Child. (KAR P3)

There is a need for setting up a Northern Region Committee on the Rights of the Child (CHA P8)

There should be regional work plan for achieving the rights of the child. (CHA P8)

Government, NGOs and community should work together to grant the children their childhood. (B1H P28)

Benefits of programmes like vocational training, tribal hostel or Ashram facilities, non-formal education centres, etc. should reach the beneficiaries.

Meaningful efforts should be made to remove any discrepancies in such schemes. In case the schemes need modifications, NGOs and Government officers/ employees should propose reforms in an organised manner to ensure successful implementation of schemes and enjoyment of rights and opportunities by children. (MP P9)

Panchayats and other NGOs should play a prominent role in publicising rights of children. (MP P11)

Basic needs of children should be identified and measures taken to meet them. (RAJ P33)

A timetable of activities for an year should be chalked out and also the use of all methods including marches, exhibitions, slogans, puppet shows. (RAJ P33)

24. Policy and Planning

There is a need for induction of action groups in policy and planning process. (CHA P8)

Representatives of NGOs, experts and regional functionaries from all levels should be involved in the formulation and evaluation of state plans. (UP P47)

25. Pre-school Education

More number of pre-primary centres should be established in the rural areas. The criteria for establishing anganwadi centres should be further relaxed. One centre for every 30 children should be started. (AP P18)

NGOs should be encouraged to start more pre-schools. (AP P18).

26. Recreation

Children's film festivals and film shows should be organised regularly for their entertainment (WB P4)

Recreation should be given equal importance in schools in addition to the academic requirements. (BIH P25)

Arrangements should be made for recreation centres, libraries and museums, etc. for children in each area which would increase their knowledge, provide entertainment and develop their mental faculties and talents. (UP P4S)

It should be ensured that proper films and television programmes are shown which leave a positive impression on the child's mind. Vulgarity on public media should be banned. (UP P45)

Holidays should be used productively by organising education tours, SUPW, shramdans in rural areas, etc. (AP P20)

Extra-curricular activities should be identified and made compulsory in schools. Scouts and guides programme and other programmes may be introduced. (AP 20)

27. *Role of Media*

There is a need for increased awareness and sensitisation campaign on child rights through media for reaching out to people. Media must play an active role in promoting CRC. NGOs and the media must interact positively to highlight violations of children's rights. (CHA P8) (WB P5) (KAR P4) (MAH P8)

Mass media must play a key role in educating parents and children and make them aware of the importance of education. (AP P13)

Haats (village markets), fairs and festivals should be used effectively for awareness creation and dissemination of messages. (BIH P23)

Success stories should be highlighted through media for recognition and motivation. (BIH P25)

Experiences may be shared regarding what worked and what failed. (RAJ P33)

Parents and children should be educated and informed about rights of children. A simplified version of CRC for children should be prepared and disseminated. (BIH P28) (MAH P7)

Director AIR Patna suggested that NGOs may organise group listening at village level on specified programmes on CRC and give regular feedback to AIR. (BIH P30) (MAH P7)

Appropriate communication materials for information and education should be developed and duplicated in large numbers for field use. This includes scripts for drama, songs, posters, hand bills and portable exhibition kits. Government/and UNICEF should jointly develop these materials and NGOs should use them in the field. (BIH P30)

Animators and volunteers are to be trained in effective communication for organising exhibitions, communication campaigns and field publicity activities. (BIH P30)

Monitoring and evaluation should be in-built in the system to study the usefulness and impact of these efforts in communication. (BIH P30)

NGOs should win the confidence of the community and make the public more aware of the rights of children using media like audio-visual aids, films, radio, theatre, folklore, folk dance, street plays, puppet shows, etc. to get rid of the evil of child labour. (MP P9)

Information regarding all the provisions related to rights of children should be disseminated through all available media, specially to organisations working in the field of child welfare such as Anganwadi Workers Training Centres, Bal Sevika Training Centres, Montessori Training Centres,

Basic Training Centres and Non-formal Education Centres. (MP P11)

Radio

(i) Regular contact should be maintained by NGOs and other Government institutions with AIR stations to broadcast the message of child's rights through plays, folklore, stories, talks in local dialect. (MP P10)

(ii) Time slots allotted for these programmes should be such that children, parents and the public may get benefit of the programme. (MP P10)

(iii) Thoughts, experiences and interviews of children should go on the air through short stories, plays, dialogues, jokes, etc. (MP P10)

(iv) Prior intimation regarding these programmes should be given through radio and newspapers. (MP P10)

(v) The villagers should be encouraged through Panchayats to listen to such radio programmes. (MP P10)

Television

(i) Time slot should be such that children, their parents and other viewers in large numbers get to see such programmes. (MP P10)

(ii) Prior intimation of programmes related to rights of children should be given from time to time, specially in between popular programmes such as Teer Kaman, Bal Jagat, Chitrahar, Film Shows, etc. (MP P10)

(iii) Information concerning the rights of children should be disseminated through programmes like cartoon films, puppet shows, plays, folklore, folk- dances, interviews, discussions, etc. (MP P10)

(iv) Adequate time and coverage should be given to news concerning children and their rights. (MP P10)

Newspapers

(i) Outstation programmes for editors and correspondents of newspapers should be organised so that they can carry out their responsibilities in a proper way by publishing prominently news related to child labour and rights of the child in the newspaper. (MP P10)

(ii) In addition to their regular columns on children like Phulwari, Bacchon ki Duniya, articles, stories, Nanhi Duniya, etc. newspapers should also accord priority and importance to publication of articles on child labour and related problems. (MP P10)

(iii) In order to create awareness amongst the public they should carry news related to schemes for beneficiaries and important events in an attractive manner. (MP P10)

Local Programmes

Information relating to rights of children should be disseminated by organising programmes like street plays, puppet shows, seminars, theatre, dance, etc. at various places through Kala Mandalis. Children fairs, games, cultural programmes, debate competitions should also be organised. Exhibitions, children campaigns, parent campaigns and rallies, should be organised to oppose injustice being meted out to children. (MP P10)

The media can play a pivotal role in disseminating awareness about development issues in general and children's rights in particular, (WB P3)

The media also has a very important role in monitoring the implementation of programmes related to the rights of the child. (WB P3) (MAH P6)

NGOs should develop a Special Media Cell to enhance the above process of monitoring implementation of children's rights. Such Cells should be headed by a Coordinator who will be responsible to ensure mutual cooperation and dialogue between the media and the NGOs. (WB P3)

The media may play a 'supportive' role complementing the efforts of NGOs, but the onus is on the NGOs to ensure the media give them enough coverage. As the audio visual medium is usually a more effective vehicle for carrying public messages more effort should be given for their utilisation. (WB P3)

Sensitisation of the media too is required. For this purpose orientation of journalists till the district level should be planned. (WB P3) (MAH P6)

For children to have access to information on their rights, they should be addressed through the school curriculum. Child-to-child communication among children can help them transmit knowledge to their peers. (WB PS)

Schools, anganwadis, children's organisations, NGOs and mothers' organisations should be approached with Information, Education and Communication packages on the Rights of the Child. (WB P5)

Awareness may be created in the community using audio visual and other methods e.g. Prabhat Pheri, Sandhya Pheri, Theatre, Street Plays, Puppet Shows, Cultural Shows, New Games, TV Channels, Video Films, Audio Cassettes. Cultural aspects should be kept in view. (RAJ P31)

Workshops should be organised on rights of the child. Girls should specifically be involved and Bal Mela and Children's Day could be effectively organised. (RAJ P31)

Schools and parents should be contacted and made

aware of the effect of our altitude towards children. (RAJ P32)

To increase awareness in the community, a march or Padyatra could be held. (RAJ P32) Slogans on child's rights should be made and written all over in the villages. (RAJ P32)

A Resource Centre should be created where there are books and material on Rights of the Child. (RAJ P32)

There should be a time slot on radio and TV when children could be called and talks may be given. Culturally oriented songs may be used for messages. (RAJ P33)

UNICEF should produce material on mass scale e.g. posters (designed by Safdar Hashmi), songs, slogans. (RAJ P33)

Newspaper should be contacted for giving wide publicity to the rights of the child. (RAJ P33)

Media should highlight those experiences which are positive in any community e.g. in some tribal areas, gender discrimination is less. (RAJ P33)

The print media are still very important and must be cultivated to increase awareness. The vernacular media must be given greater importance as their influence is large and continues to grow. (MAH P6)

28. *Role of NGOs*

NGOs should function as pressure groups in improving the quality of services by creating a demand for them and by motivating the community for active participation (BIH P23)

NGOs have a key role to play in planning, implementation, monitoring and evaluation of the process of implementing the rights of the child. The state level task force on CRC implementation already formed should have adequate NGO representation. (BIH P19) (UP P47) (AP P23)

NGOs should send their recommendations to the task force which is set up to look into the discrepancies that exist between the Constitution and the Convention. (KAR P4)

To protect the rights of the child, NGOs should undertake the following:

(i) Raise public awareness on various national and international laws on Rights of the Child. (ORI P6)

(ii) Protest against the policies and programmes hampering the interests of the children. (ORI P6)

(iii) Undertake advocacy and lobbying to defend the rights of the child. (ORI P6)

(iv) Undertake welfare activities for specific target group among the children. (ORI P6)

Interaction with local NGOs to highlight the situation of children in institutions is important, (AP P17)

NGOs should organise awareness and training programmes for policy makers as well as individuals at the grassroot level. (WB P4)

Periodic monitoring of programmes should be done through NGO networks and other systems of assessment. (WB PS)

NGOs should provide more and better information to the media. They should be oriented on developing competence in communication, i.e. on giving items of news value to the media. (MAH P6, 8).

The UN Convention on the Rights of the Child came into force as an international law on 2 September 1990. The Government of India acceded to and ratified the Convention on 12 November 1992. The Convention comprising 54 articles, is a set of international standards and measures that intends to protect and promote the wellbeing of children in society.

It provides the legal basis for initiating action to ensure the children their rights related to survival, protection, development and participation. The World Summit on Children held on 30 September 1990, strongly endorsed the Convention. Infact, this was the first major global action for the implementation of the Convention. India being a signatory to the World Declaration, has formulated the National Plan of Action for Children, which provides a framework for actualisation of the Convention and a practical agenda to guarantee the rights of the child set forth in the Convention.

Today in India, the situation of children has improved considerably. A child who was earlier considered as dependent in every respect on parents is now emerging with his/her own independent identity and is entitled to rights, including self articulation of development paths and goals. Existing constitutional provisions, policy framework, legislation and above all the commitment at the highest government level gives an impetus to the efforts being made to meet the basic needs of the children.

As per UNICEF's State of World Children Report 1995, India has made dramatic progress in protecting the health of children and promoting immunisation and iodisation programmes for them. Still, all the children in the country do not have equal opportunities and facilities for living and learning. They do not have the same level of social acceptability. About 19-24 million children in the age group 6-14 years are illiterate and only 52 per cent of the enrolled children reach primary school. Malnutrition is still a major area of concern. As per Operational Research Group study (1983), there are 44 million child workers, some are bonded and some are sold in prostitution.

Further, it is estimated that there are about 500,000 street children in seven cities of India. Children living in difficult circumstances like street children, destitutes, juveniles, children of prisoners, prostitutes, etc. are nearly deprived of the basic needs. Due to existing social problems like infanticide, child labour, child prostitution, dowry, gender inequality, child marriage etc. many children in the country are missing out on their childhood. It is hoped that with the implementation of the Convention, this may not be the reality in the near future. The Convention on the Rights of the Child is a landmark document, at the national and international level, which has facilitated conversion of the basic needs of the children into their normative rights. It is as important for tackling the children as the Constitution is for the country. In a sense, it is a means of empowering children and creating an environment in which all children are able to live securely and realise their full potential. While the Convention obligates the Government to ensure observance of its provisions, the responsibility of implementing the Convention lies with all concerned with children. Infact, all the Articles of the Convention do not require the intervention of judiciary or Government delivery system to give the children their rights.

Most of them are social and cultural commitments, where the community and the parents have a very important role to play. Therefore, in order to arouse public consciousness and political will in support of the child's rights, it is all the more necessary to raise awareness about the Convention and the provisions made therein. Article *42* of the Convention states that States Parties undertake to make the principles and provisions of the Convention widely known, by appropriate and active means, to adults and children alike. However, it is felt that the awareness about the rights of the child among children and adults is very limited.

It is reported (Maheshwar, 1994) that the percentage of children who assert their rights is negligible and 30 per cent of the children are not even aware of their fundamental rights due to lack of education. Similarly, amongst the adults, 10 per cent are aware of the rights of the child, 15 per cent give this subject a thought and only 7 per cent are engaged in facilitating these rights to them.

The most important task before the government is to initiate a plan for creating awareness and disseminating information contained in the Convention to every citizen of the country in the manner in which he can understand the children's rights as a judicial law. People from various walks of life i.e. policy makers, administrators, teachers, lawyers, professionals, medical practitioners, editors, journalists, social workers, opinion leaders, folk artists, etc. must know the details of the Convention and its implications, so that they can willingly participate in the development programmes for children. The most appropriate approach for creating awareness and wide dissemination of information about the Convention across the whole section of the society is communication media. Communication media can be a powerful advocate and a tool for awareness creation for the rights of the child. Wilber Schramm is of the opinion that media has the power to initiate, stimulate, provoke and promote social development by providing varied perception, interest, motivations, beliefs, faiths and ideas. It is the media which mediates in the production, consumption and accumulation of goods and services, besides promoting potential, social, psychological and cultural processes. Communication field has a magnetic effect, and any information regarding the rights of the child disseminated through the media will be readily received by children and adults, provided it is presented in an interesting manner to hold their attention and is delivered at the right time. So the role of media in promoting the rights of the child cannot be undermined.

The role of media in implementing the Convention can be envisaged as that of advocacy and social mobilisation to empower all concerned with children, by creating awareness and transferring knowledge, skills and resources related to the basic needs and rights of the child, so as to promote participatory development and adopt measures which give effect to the rights recognised. As per the Convention, the major areas of concern related to children which need media support are health, nutrition, education, family environment, gender inequality, child labour, welfare of children in difficult circumstances, etc.

Accordingly, the media interventions for advocacy and awareness creation of the rights of the child are:—

(i) sensitising administrators, NGOs, police, judiciary, professionals, social workers and the community about the children's rights;

(ii) reminding government of the commitments made;

(iii) informing people of the existing legal provisions, spreading legal literacy and explaining the laws as well as structures and methods of administration of justice in a simplified way;

(iv) educating parents on the rights of the child and their role.

(v) alerting public of the changing situation of children and women and the services available for their welfare and development;

(vi) bridging information and awareness gaps about the issues related to child's rights to bring about desirable change in the knowledge, attitude and practices of the community.

(vii) monitoring and protecting the child's rights;

(viii) lobbying the competent bodies and persons to take corrective action for rights violated;

(ix) making children understand the concept of rights, to know what rights they are entitled to, and to empower them to participate and take action to enjoy these rights.

(x) promoting democratic process of participatory development and commitment for ensuring the children their rights.

At present, the use of media for the child's rights is very limited. Health related issues of national importance get coverage on electronic media at the prime time. Of late, child labour and gender inequality have caught the attention of media professionals. To use media effectively to promote rights of the child, it is essential to consider the pre-requisites for the use of media like communication infrastructure, communication needs, awareness among media professionals, journalists, editors and folk artists, availability of communication software for all media forms, trained manpower, a system to monitor and supply items of news value to the media, coordination and integration among all concerned with the implementation of child's rights for use of media, and above all the political commitment in favour of media.

If we take stock of the communication infrastructure in the country, we will find that the communication system in the country has steadily expanded and has become one of the largest in the world. Today, the radio reaches 96.2 per cent of the population and the television has a reach of 82.4 per cent. The films produced by the Films Division have an audience of about 5 to 6 crores every week. There are about 30,214 newspapers with circulation of nearly *539* lakh copies in 16 major languages and dialects. This infrastructure comprising various communication channels and media can be used for disseminating practical, target-oriented information about the rights of the child.

Though using this vast communication infrastructure to create awareness and to bring about behavioural change on issues related to child's rights is not an easy task, yet it can be accomplished with proper planning, coordination, and by formulating area-specific and need-based action programmes. Infact, there is a need for a Forum or an Action group to facilitate the use of media for the rights of the child by evolving timebound media strategies for implementing and networking communication activities.

There is a wide variety in traditional and modern media and non-formal communication techniques available. Mass media channels are being used for creating awareness, and imparting knowledge and information, and group and interpersonal communication techniques are used for decision making and behaviour change. Electronic, print and folk media are being extensively used both for entertainment and education. Electronic and print media have better reach in urban areas. Folk media is more acceptable at the grassroot levels, and is being used in remote areas where mass media technology has not fully reached.

All these media forms have their strengths and limitations and differ in terms of credibility, physical and psychological accessibility, reach, and other characteristics. Different messages related to child's rights may call for the use of different media, depending on characteristics of various media, there availability, reach and the target audience. On the whole, communication in support of rights of the child requires an integrated deployment and use of media-mix approach and social marketing techniques.

As per Article 17 of the Convention, mass media has a wider role to play in promoting the social, spiritual and moral well being of children. The Article clearly states that States Parties recognise the important function performed by the mass media and shall ensure that the child has access to information and material from a diversity of national and international sources, especially those aimed at the promotion of his or her social, spiritual and moral well-being and physical and mental health.

Now-a-days the popular mass media forms are electronic and print media. Radio and television, soap operas, prime time television programmes, radio jingles and spots, TV quickies, motion pictures, comics, cartoons, books and popular music are being used for disseminating educational messages related to health and family welfare. The same software of electronic and print media can also be used for creating awareness about rights of the child. At the community level, formation of listnen groups/ radio forums can give regular feedback to AIR/Doordarshan on issues related to child's rights. The producers and directors of AIR and Doordarshan need to be encouraged to produce programmes for children with focus on the rights of the child in which the children can also participate.

Children should be provided with a opportunities to express their feelings and opinion using plays, skits, stories, talks, debates, etc. It is important that these programmes are allotted suitable time slots so that both, children and adults can benefit out of it. Print media especially the local press can play a key role in publicising the Convention among the literate group.

Appropriate target-specific print material has to be produced on all aspects of child's rights for children as well as adults. Infact, the simplified version of the Convention should be prepared and translated in various regional languages for children to understand it Various forms of print material like posters, charts, flash cards, flipbook, etc. are needed for effective interpersonal communication to promote child's rights.

Traditional/folk media are presently being used by Song and Drama Division and Field Publicity Units of the Ministry of Information and Broadcasting for disseminating health-related messages at the grassroot level. The potential of this media can also be explored for promoting development issues concerning rights of the child in the community. Further, fairs, festivals and haats (village market) can also be used as a forum for creating awareness and disseminating messages related to child's rights.

In our country, the Government controls the principle channels of communication i.e. radio and television. The importance accorded for advocacy of child's rights on these channels depends largely on the priority the government places on child's rights as well as the attitude of media professionals. To harness the public media for rights of the child, it is necessary not only to sensitise policy makers and planners concerned with child welfare and development, but also the media professionals, editors, journalists, folk artists, etc. Joint consultative meets, workshops and orientation courses for media and subject experts can provide a forum for creating awareness and exchanging information about child's rights.

Both electronic and print media can also monitor the progress made in achieving the realisation of the obligations undertaken in the Convention. Infact, media can become the watchdog to protect children's rights by highlighting success stories as well as cases of violation of rights. It is assumed that once the media starts taking interest in the rights of the child, the Government will also start taking the Convention more seriously.

The Government of India is making earnest efforts to create awareness about the rights of the child. During 1994, the theme selected to commemorate the Universal Children's Day on 14 November was, Rights of the Child: A Commitment. On this occasion the Department of Women and Child Development released a document giving in brief, the historical perspective of the rights of the child and situation of children with reference to articles of the Convention in the spheres of health, nutrition, education and related aspects of social support. All the State/UT Departments were instructed to make concerted efforts for advocacy for child's rights on the basis of guidelines issued. In addition, simple and selected messages were developed by the Department on rights related to survival, protection, development and participation of children. These messages received adequate coverage in the leading national newspapers.

During the year, UNICEF in collaboration with the Department of Women and Child Development, ICCW and its State branches, and selected NGOs organised eleven workshops at the national and state level to create awareness about the Convention among NGOs and also to identify the areas for promoting the implementation of the Convention with their support.

The Department of Women and Child Development and UNICEF have also been working with the Ministry of Information and Broadcasting in a number of States to bring child's rights into the forefront. An action plan for publicising the Convention on the Rights of the Child has been prepared and is being implemented. The major activities being proposed under the Plan of Action include orientation/ training of staff, development of print material, and programmes for children on radio and TV.

Making child's rights a reality is a joint effort. NGOs can join hands with the Government to make this dream come true. NGOs being nearer to the masses are in a better position to organise the community for ensuring the children their rights. These organisations can play an important role in promoting use of media for advocacy and awareness creation for the rights of the child. Some of the tasks that can be undertaken by NGOs in this direction are as given below.

(i) NGOs can interact positively with media professionals from all fields to highlight success stories as well as violation of child's rights. They should be oriented on developing competence in giving items of news value to the media.

(ii) NGOs at the grassroots level can become effective communicators or change ageats*, and can mobilise and empower the community for their active participation and commitment to ensure the children their rights.

(iii) At the community level, they can also mobilise social and cultural organisations, recreational groups, mahila mandals, etc. for child's rights.

(iv) There is a dearth of communication software for both electronic and print media. NGOs can be actively involved in planning and production of software and messages.

Already the Government of India has made efforts to sensitise NGOs on issues related to child's rights. Now, initiatives have to be taken by NGOs to introduce communication interventions and use the power of media to realise the obligations undertaken by the Convention.

To sum up, the Convention on the Rights of the Child is a landmark document and is the first legally binding global prescription to give to the children the best they deserve. Hence, creating awareness about the Convention, its application and monitoring is a necessity and calls for a joint effort of all concerned with children. Communication media can be a powerful tool for advocacy and awareness creation provided it is used in a planned manner. At present, the use of media is limited for the child's rights. There is a need to set up an Action Group on Media, to promote sustained communication efforts on the issues related to child's rights, sensitise media professionals, train communicators, develop adequate messages and software for electronic and print media, use traditional media effectively, involve children in advocacy programmes and coordinate with all the Government Departments, NGOs, professionals, subject experts, etc. working for children. NGOs can play an important role in the use of media.

REFERENCES

Bihar State Council for Child Welfare. The Convention on the Rights of the Child: A State Level Workshop for NGOs in Bihar, 1994, Oct. 4-6, Walmi. Patna, the author, 1994.

Dua, M. R. and Gupta V. S. (Ed.) *Media and Development: Theories in Communication and Extension*. New Delhi, Har-Anand Publication, 1994.

India, Ministry of Human Resource Development, Department of Women and Child Development. Universal Children's Day 14 Nov. 1994, Rights of the Child: A Commitment. New Delhi, the author, 1994.

India, Ministry of Information and Broadcasting. *India 1993: A Reference Manual.* New Delhi, Ministry of Information and Broadcasting, Publication Division, 1994.

Maheshwar, Madan Lal. Role of Media in Creating Awareness for Children Rights in *A Report of the Workshop on the Rights of the Child,* 1994 August 30-31, Jabalpur, M.P.

Mahila Dakshata Samithi. *State Level Workshop on Rights of the Child: A Report.* 1994, Sep. *27-28,* Hyderabad. Hyderabad.

Mathur, Kanwar B. *Communication for Development and Social Change.* New Delhi, Allied Publishers, 1994.

UNICEF. *Rights of the Child.* New Delhi, the author, *1992.*

UNICEF. *The State of the World's Children,* 1995. New Delhi, the author, 1994.

Windahl, Sven *et al. Using Communication Theory: An Introduction to Planned Communication.* London, Sage Publications 1992.

8

Towards a Society for All

It is pertinent to note that the global crusade against the vicious triad of malnutrition, morbidity and mortality has often underplayed a related dimension—the disability dimension. The world's threshold of tolerance towards disability remains much too high. The time has come for a global offensive on the disability front— every bit as powerful as the one being waged against mortality. Despite the global commitment to human rights and equal opportunities, preventable barriers to the achievement of both individual development potential and societal human resource development persist. These barriers include micronutrient deficiencies and vaccine preventable diseases, among a spectum of other factors which can be prevented and controlled on a large scale, using available low cost technologies. Action for prevention and rehabilitation of disability is not only a necessary investment in national human resource development, productivity and economic development—but the right of each individual.

In recent years the World Development Report 1993 signified the attempt to highlight the disability dimension, by evolving a unit of measurement called "Disability Adjusted Life Years", while examining human resource development. Globally efforts for disability prevention and rehabilitation were intensified with the formulation of World Programme of Action concerning Disabled Persons, adopted by the United

Nations General Assembly in 1982. The period 1983-1992 was proclaimed as the United Nations Decade of Disabled Persons. This decade witnessed the progression from awareness raising to action oriented measures, aimed at the continued improvement of the situation of persons with disabilities, and the equalisation of opportunities for them. In 1990 the clear focus on action crystallised, with the aim of achieving a Society for All by the year 2010.

Despite the appreciable increase in activities designed to enhance public awareness of the needs and circumstances of people with disabilities and related issues, the need for sustained efforts to overcome physical and social barriers to the full equality and participation of disabled persons continued. Aware of the need for more vigorous and broader action and measures at all levels to fulfil the objectives of the Decade and the World Programme of Action, on October 14, 1992, the UN General Assembly called for the observance of December 3 as the International Day of Disabled Persons. The observance of the International Day of Disabled Persons is a unique opportunity to focus attention on the World Programme of Action, for promoting effective measures for the prevention of disability, for rehabilitation and the realisation of the goals of full participation of disabled persons in social life and development and equality of opportunity. The significance of this day is marked by its contribution in awakening consciousness regarding the gains to individuals and society from the integration of persons with disabilities in every area of social, economic and political life.

Perhaps on the International Day of Disabled Persons—3 December, 1994 as we move towards a Society for All by 2010, we can make a new beginning:

For Disabled—The Differently Abled

- Inadequate care and micronutrient deficiencies during pregnancy, birth asphyxia caused by unsafe delivery practices, and vaccine preventable diseases such as polio, tuberculosis and measles present a major risk of disability.

Birth asphyxia results in infant deaths as well as cerebral palsy, and related disorders in surviving infants. Polio is an infectious crippling disease. Tuberculosis is a chronic debilitating disease, affecting different parts of the body. Measles can also result in secondary infections. This includes middle ear infection—which can cause hearing impairment.

Promote safe motherhood, universal child immunisation, safe drinking water and environmental sanitation, to prevent childhood disability

- Vitamin A is essential to growth and to the development of immunity to diseases, especially among young children. An estimated 5-7 per cent children in India suffer from eye damage caused by vitamin A deficiency. While the short term intervention is focused on administration of mega doses of vitamin A on periodic basis, dietary improvements is the long-term ultimate solution to the problem of vitamin A deficiency.

Promote exclusive breast-feeding in infants (upto 4-6 months of age)—including feeding of colostrum; consumption of vitamin A rich foods; and vitamin A supplementation of children as per recommended schedule.

- Deficiency of iodine is the world's single most significant cause of preventable brain damage and mental retardation. Iodine deficiency results in lowering the average intelligence of the entire school age population by as much as 10 to 15 IQ points per child. Daily consumption of iodised salt prevents these disorders caused due to iodine deficiency. *Consume only iodised salt.*

- The first level or primary prevention of childhood disability addresses causes of disabling conditions. The second or next level is prevention of disabilities resulting from these conditions or diseases, when they occur.

  Almost three quarters of common impairments can be prevented from deteriorating into disability—through care, early detection of developmental delays, stimulation and intervention. Early action is important.

Promote capacity—building of care givers—parents, siblings and childcare workers. Provide a nurturing and stimulating environment to the young child—through parent-child, child to child and workerchild interaction. This will help promote early childhood care and holistic development of the young child

- The third level is tertiary prevention or alienation of the effects of disruption of the normal processes of child development, when disability has occurred. This prevents disability from becoming a handicap.

  The child has the right to opportunities to achieve full development potential. The disabled child has the right to care, protection, integration and participation in the development process. The learning needs of persons, (especially children) with disabilities—demand special attention, to achieve both individual development potential—and the national commitment to "Education for All".

Promote the rights of the child—including the rights of children with special needs

- Prevention of and protection from childhood disability is the right of the child and our responsibility. We all have a role in promoting child development the foundation of human resource development.

Let us work together—people's representatives, policy planners, opinion leaders, decision makers, programme managers, frontline workers, representatives of voluntary agencies, social activists, community leaders, communities and families to prevent disability arising from causes we know how to prevent.

And let us strive to enable the differently able child by empowering communities and families with the necessary knowledge and attitudes for the care, development, protection and participation of the "One in Ten " child

Government Initiative Government of India National Plan of Action a Commitment to the Child 1992 Illustrative Lis' of Goals for the Year 2000

It is estimated that one in every ten children in the world is born with or acquires a physical, sensory or mental disability caused by disease, malnutrition, environmental factors or accidents. Approximately three quarters of disabilities result from causes we know how to prevent.

Many of the direct and underlying causes of childhood disability are those addressed in the World Declaration on the Survival Protection and Development of Children 1990 with clearly defined goals for the year 2000 A.D. In India the national commitment to the achievement of these goals for children is reiterated in the National Plan of Action for Children 1992. These goals for the year 2000 A.D. are:

Survival

- Reduction of IMR to less than 60 per 1,000 live births,
- Reduction of U5MR to less than 70 by 2000 A.D.,
- Reduction of matenial mortality rate by half.

Health

- Eradication of poliomyelitis by the year 2000.

- Elimination of neonatal tetanus by 1995.
- Reduction by 95 per cent in measles deaths and reduction by 90 per cent of measles cases compared to pre-immunisation levels by 1995.
- Achievement and maintenance of high level of immunisation coverage at a level of 100 per cent of infants.
- Reduction by 50 per cent in deaths due to diarrhoea in children under the age of 5 years.
- Reduce mortality rates due to ARI among children under 5 by 40 per cent by 2000 A.D. from the present level.

Nutrition

- Reduction in severe and moderate malnutrition among under-5 children by half between 1990 and the 2000.

Education

- Universal access to primary education with special emphasis for girls and accelerated literacy programme for women.

Water and Sanitation

- Universal access to safe drinking water and improved access to sanitory means of excreta disposal

Children in Especially Difficult Circumstances

- Improved protection of children in especially difficult circumstances.

The Girl Child

- To remove the gender bias and to improve the status of the girl child in society as to provide her with equal opportunities for her survival and development to her full potential.

Children and the Environment

- To conserve and protect environment so that it is conducive to the health and well being of children.

Advocacy and People's Participation

- Advocacy for the child as everyone's concern.

Achievement of major goals for promoting child health, nutrition, education and access to water and sanitation will contribute substantively to primary prevention of childhood disability by addressing causes of disabling conditions, diseases and impairments. A spectrum of national programmes are being implemented for primary health care, child survival and safe motherhood, universal immunisation, Integrated Child (Development Services, prevention and control of micronutrient deficiency, safe drinking water and environmental sanitation and education to achieve these goals for children.

The Right of the child to survival, development, protection and participation is embodied in the UN Convention on The Rights of The Child, and ratified by the Government of India. The convention also enshrines the right of the children with disabilities to be protected against the effects of disruption of the normal processes of child development. The Ministry of Welfare in partnership with different allies and partners is actively promoting the prevention, early detection, intervention and community based rehabilitation of childhood disability.

Reaching the Differently Able

Attitudes

- There is an history of negative messages about people with disabilities that can be identified and changed.
- Children and adults with disabilities, or their families are our best spokespersons and advisors.

- The attitude portrayed in our prevention messages can either be de-humanising or humanising.
- There are guidelines which can be followed when selecting messages and designing product; which will help to promote positive and sensitive attitudes about disability.

Early Detection

- Many disabilities can be detected at birth or within the first year of a child's life.
- The earlier a disability is detected in any child, the more effective is the intervention and often, the less severe is the disability.
- Early detection has been and can become part of all existing Primary Health Care and Early Childhood Development Activities.
- People need not be experts or highly trained to detect disabilities in young children.
- Material that can be found in every village or city can be used to detect a disability.

Interventions

- Intervention with a child with disabilities should begin as early as possible.
- Intervention can and should take place, whenever possible, within the family and community systems.
- Rehabilitation interventions need not be sophisticated, expensive or carried out solely by specialists.
- Children with disabilities should interact with non-disabled children in as many aspects of their lives as possible.
- The Skills children learn during the ages of 0-6 years prepare them for living independently.

Community—Based Rehabilitation (CBR) a Step towards Independence

- The rehabilitation interventions made with a child who is disabled should be those that make sense to and empower the child and her/his family.
- Children (and adults) with disabilities are doing and can do things that many people thought impossible.

Family and Community Involvement in CBR

- To the extent possible, children who are disabled and their families should be involved in all decisions regarding their rehabilitation.
- Low-cost, already available materials and existing person-power within communities are often all that is necessary for effective CBR.
- Most of the activities that are beneficial to children who are disabled are equally beneficial to non-disabled children.
- Families are often fearful of rehabilitation interventions and some tend to over protect their child who is disabled.
- Families need time to be convinced of the need for CBR and the fact that they are the best "rehab workers."
- A CBR worker needs to slowly demonstrate activities to family members and then guide them through the same hands-on activities.
- Basic stimulation of a child's senses is the basis of many rehabilitation activities.

Child to Child

- Young children can and do have an impact on their communities.
- Young children can have an impact on the lives of children with disabilities.

- Activities that are beneficial to children with disabilities can also improve the skills of the non-disabled children who are conducting the activities.
- When demonstrating a need for some change to benefit a person who is disabled, try to show how *many non-disabled people* will also benefit from the change.

Programmes/Schemes of the Ministry of Welfare (HW Division) Ministry of Welfare, Government of India

Estimates of the number of disabled vary, depending on the definitions, the methodology and the extent of use of scientific instruments in identifying and measuring the degree of disability. The National Sample Survey Organisation (NSSO) conducted in 1981 a countrywide survey covering 3 types of disabilities, i.e., visual, communication and locomotor disabilities. It identified 12 million persons having atleast one or the other disability-constituting about 1.8 per cent of the estimated total population. In the year 1991, NSSO again conducted a (countrywide survey covering disabilities:

(i) visual disability,
(ii) hearing disability,
(iii) speech disability and
(iv) locomotor disability.

About 16.15 million people are estimated to have one or the other of the 4 types of disabilities mentioned above. This constitutes 1.9 per cent of the total estimated population of the country. It has been observed that for the country as a whole the prevalence of physical disability was 20 per thousand persons in rural sector and 16 in urban sector. Between the two sexes, the prevalence is marginally more among males than among females.

A sample survey conducted by NSSO in 1991 for persons with delayed development between 0-14 age estimates that about 3 per cent of the estimated child population are having delayed development. However, on the basis of some random sample surveys, it has been estimated that about 2 to 2.5 per cent of the population of the country is mentally retarded.

According to National Programme for Control of Blindness (WHO Report 1989) about 28.56 million persons are with low vision. A study conducted by Indian Council for Medical Research estimated that 6.8 per cent people in urban areas and 10.8 per cent people in rural areas has significant hearing losses.

The number of leprosy affected persons is estimated to be about 4 million, of whom l/5th are children. About 15 to 20 per cent cases are with deformities. In 196 districts in the country, the prevalence rate is more man 5 per thousand. About 430 million persons live in these high endemic districts.

Since the advent of independence, the Government of India have taken a number of initiatives in the field of special education, vocational training, rehabilitation, manpower development and technology upgradation. Important Schemes of the Ministry are as follows:

Assistance to Voluntary Organisations for the Disabled

The Ministry of Welfare, Government of India gives upto 90 per cent assistance to voluntary organisations for the education, training and rehabilitation of the disabled. For rural areas, assistance is given upto 95 per cent. In 1993-94 about 315 voluntary organisations were provided with assistance to the tune of Rs. 10.40 crores. A component to rehabilitate mentally ill persons has been added in the scheme.

Assistance for Aids and Appliances

The Government of India under this scheme provides free aids and appliances to those persons with disability whose monthly income does not exceed Rs. 1200 while 50 per cent subsidy is provided to those whose income is between Rs, 1201-2500 per month. This scheme is being implemented through both governmental and non-governmental agencies.

Assistance to Voluntary Organisations for Rehabilitation of Leprosy-Cured Persons

This scheme envisages providing financial assistance to voluntary organisations working for leprosy-cured persons. Assistance is given upto 90 per cent to such voluntary organisations who develop programmes for awareness generation, early intervention, educational and vocational training, economic rehabilitation and social integration of the leprosy-cured persons.

Assistance to Voluntary Organisations for Persons with Cerebral Palsy and Mental Retardation

The scheme aims at developing organisational and infrastructural facilities for manpower training and professionals, hostels and other assistance required for imparting training of various categories of workers/trainers such as vocational teachers, rehabilitation workers, wardens, etc., in the field of Cerebral Palsy and Mental Retardation. Assistance is given upto 100 per cent of expenditure for recurring and non-recurring items.

Assistance to Voluntary Organisations for Establishment of Special Schools

Under this scheme, grant is given to voluntary organisations upto the extent of 90 per cent for the establishment of Special Schools. Preference is given for opening schools in new districts and upgradation of existing schools. The scheme was started in 1993-94. The ultimate objective of the scheme is to establish at least one special school in each district of the country with the help of the voluntary sector.

National Trust for the Welfare of Persons with Mental Retardation and Cerebral Palsy

A National Trust for the Welfare of Persons with Mental

Retardation and Cerebral Palsy is proposed to be set-up. An amount of Rs. 1.25 crores has been ear-marked for this purpose.

The objectives of the proposed National Trust would be as follows :

(i) To arrange to provide care and rehabilitation to the mentally retarded and the cerebral palsied as a step towards social security who are under the guardianship of the Trust.

(ii) To lay down guidelines for the improvement of the existing organisations which are engaged in taking care of mentally retarded and cerebral palsed depending upon the availability of finances.

(iii) To set up homes and service institutions for providing residential care to persons with mental retardation and cerebral palsy.

(iv) To provide financial and technical assistance to the organisation providing care and rehabilitation services to persons with mental retardation and cerebral palsy.

(v) To provide guardianship and foster care and to take over the guardianship rights of the persons with mental retardation and cerebral palsy after the death of the parents, if so, desired by the family or in absence of family support.

(vi) To extend and support the welfare programmes of families/foster families/parent associations and voluntary organisations.

(vii) To provide legal aid to the mentally retarded persons and their families.

(viii) To receive, own and manage properties bequeathed by the parents to maintain their child with Mental Retardation or Cerebral Palsy after their death.

National Programme for Rehabilitation of the Disabled

The Scheme aims at providing a more comprehensive package of rehabilitation services to the rural areas. The Scheme will take into account the following areas:

(i) It will offer rehabilitation services, viz., restorative therapy, early detection and timely intervention, parent counselling and provision of aids and appliance at the village, PHC, CHC and district levels to disabled Persons;

(ii) Provide assessment, guidance and referral facilities, for educational, vocational and placement services;

(iii) Undertake the work of awareness creation and information dissemination to tackle various aspects of disability including prevention and rehabilitation;

(iv) Rehabilitate every disabled person as far as possible in his own environment;

(v) Mobilise community support and coordinate all existing services and schemes for prevention of disability and rehabilitation of disabled persons.

The scheme is being finalised as a Central Scheme wherein, assistance would be given directly to the NGOs by the Ministry of Welfare.

National Awards in Conformity with the Unresolution the Government of India Have Decided to Observe 3rd December as International Day for the Disabled

Each year on the Occassion of the World Disabled Day, National Awards are given by the President of India to the followings:

(i) Best Employer of handicapped;

(ii) Best handicapped employee and self-employed;

(iii) Best individual working for the welfare of the handicapped;

(iv) Best institution working for the welfare of the handicapped;

(v) Placement Officers; and

(vi) National Technology Awards for Welfare of the Handicapped.

National Institutes

Four national institutes namely National Institute for Orthopaedically Handicapped, National Institute for Mentally Handicapped, National Institute for Visually Handicapped and Ali Yavar Jung National Institute for Hearing Handicapped were set up as premier institutes in their respective areas of disability to cater to the needs of the disabled in the field of education, development of manpower, training, vocational guidance, counselling, research, development of suitable service models and lowcost aids and appliances. Besides, there are two other institutes Institute for the Physically Handicapped, New Delhi and National Institute of Rehabilitation Training and Research (NIRTAR), Cuttack, Orissa which are major for the locomotor handicapped.

Artificial Limbs Manufacturing Corporation (ALIMCO)

Established in 1972 under the Companies Act with sole objective of promoting, developing, manufacturing and marketing of artificial limbs and aids and appliances, ALIMCO is the only public sector company of its type in the country. It manufactures crutches, wheelchairs, tricycles and other aids and appliances which are of international standards. There are 163 implementing agencies which provide the aids and appliances manufactured by ALIMCO.

Science and Technology Project in Mission Mode

The project was launched in 1988 with the objective to coordinate, fund and direct application of technology in the development and utilisation of suitable and cost-effective aids and appliances, methods of education and skills development leading to enhancement of opportunities for easier living, mobility, communication, recreation, employment and integration in the society 42 projects have been field tested and some like the photovolataic battery charger, interpointing braille writing frame, closed circuit TV for low-vision people and speech synthesiser have been cleared for mass production.

Media and Awareness Creation

For any people-oriented programme to succeed, the attitude of the masses towards the programme itself and towards the target-group for whom it is meant is of primary importance. In India, information about disability and the people with disability is still mostly erroneous and based on myths rather than facts. Hence the need for generating awareness amongst the masses is of paramount importance. Here we cannot overlook the pivotal role that mass media has to play. Without the help of the media-TV, radio, print, and traditional—we cannot make much headway in generating awareness. Towards achieving this objective a National Information Centre on Disability and Rehabilitation at the Centre and Four Regional Task Forces in Calcutta, Madras, Bombay and Delhi have been constituted for providing guidance and formulating time-bound programmes for creating awareness amongst the masses and for sensitising media about the portrayal of disability in a positive and humane way. The emphasis is on correct and effective messages being reflected in the media.

Employment

23 special employment exchanges and 55 special cells in regular employment exchanges are providing placement to people with disabilities, 17 Vocational Rehabilitation Centres are also providing vocational training facilities. Special employment exchanges are financed on 80-20 per cent basis and Cells are financed on 100 per cent basis by the Central Government. Central Government and Central Government Public Sector Undertakings have reserved 3 per cent posts,

one each for orthopaedically handicapped, speech and hearing handicapped, and visual handicapped. The Ministry of Welfare as the nodal agency organised special drives through the Staff Selection Commission to clear backlog for employment of disabled persons.

Rehabilitation Council of India

The Government has set up the Rehabilitation Council of India to enforce uniform standards in training of professionals in the field of rehabilitation for the disabled, Maintenance of Central Rehabilitation Registrar and other connected matters. The Rehabilitation Council of India Act has been enacted and has come inforce with effect from 31st July, 1993.

Relief Assistance Under Bilateral Agreement

Ministry of Welfare continues to operate Bilateral Agreement on gift India with the Governments of Federal Republic of Germany, Sweden, Switzerland, United Kingdom and United States of America and other U.N. organisations especially UNICEF.

Barrier-Free Environment

The physical environment including buildings, transport system, access to gardens, playgrounds, places of entertainments, etc., are designed for able bodied persons. Very little attention has so far gone into architectural design of public places and transport keeping in view the requirements of persons with disabilities. The Ministry of Welfare is in constant touch with the Ministries of Urban Development, Railways, Surface Transport, Civil Aviation and other such bodies for ensuring a time bound action programme so that a barrier-free environment with easy accessibility for persons with disabilities is promoted. For this purpose a high-level committee has been constituted by the Ministry of Welfare.

Perspective

Comprehensive prevention and rehabilitation efforts for

persons with disabilities can only succeed, with the active involvement of the Union Government, State Governments, Voluntary Organisations and society at large in cohesion. Without their concerned efforts, no prevention or rehabilitation programme can succeed.

The Ministry of Welfare has also constituted an intersectoral task force on disability prevention to network different partners in this endeavour.

Besides, the attitude of society towards people with disability has to change from patronising benevolence to that of trusting their abilities. The need of the hour is to initiate rehabilitation programmes that sustain; provide care and concern that is asting: ensure an environment that is accessible, and create awareness in the society that leads to an attitude wherein able bodied perceive what people with disabilities 'can do' and no what they 'cannot do'. It is the duty of the nation to provide people with disabilities same rights and opportunities that are taken for granted by others. People with disabilities have the right to work, the right to enjoy a full and rewarding life and the right to be integrated into society as 'one of us'. In this noble task everyone must lend his/her supporting hand.

Declaration of "Rights of the Child"

Right to Survival

- Every child has the inherent right to life.
- Timely registration of births and deaths is your duty.
- A child has his/her own identity—recognise it.
- Register my births and give me a name.
- Registration of birth and deaths are crucial for programme planning.
- Let the child enjoy the highest attainable standard of health.
- Caring for children is the prime responsibility of the parents.

- Effective and caring and child rearing gives way to preparation of responsible future parents.
- Ensure sustainability of child survival by empowering parents.
- Programming in child survival and development needs an integrated approach.

Right to Health and Nutrition

- Becoming pregnant before the age of 18, or after the age of 35, increases the health risks for both mother and child.
- The risk of death for young children is increased by about 50 per cent if the space between births is less than two years.
- Family planning gives couples the choice of when to begin having children, how many to have, how far apart to have them, and when to stop.
- The risks of child birth can be drastically reduced by going to the nearest health worker for regular check-ups during pregnancy.
- Children between the age of six months and three years should be weighed every month. If there is no weight gain for two months something is wrong.
- Breastmilk alone is the best possible food for the first four-to-six months of a child's life.
- By the age of four-to-six months, the child needs other foods in addition, to breastmilk.
- A child under three years of age needs food five or six times a day.
- A child under three years of age needs a small amount of extra fat or oil added to the family's ordinary food.
- All children need foods rich in Vitamin A.
- Intake of Vitamin A prevents blindness.
- Eat dark green leafy vegetables.
- Overcome iodine deficiency—use iodised salt.

- After an illness, a child needs extra meals to catch up on the growth lost during the illness.
- Fight poverty to eradicate malnutrition.
- Elimination of child malnutrition is an investment in human capital.
- Reduce malnutrition through appropriate interventions.
- Immunisation protects against several dangerous diseases. A child who is not immunised is more likely to become under-nourished, to become disabled and to die.
- Immunisation is urgent. All immunisations should be completed in the first year of the child's life.
- It is safe to immunise a sick child.
- Eradicate polio by 2000 AD.
- Eliminate neonatal tetanus by 2000 AD.
- Get your child vaccinated against measles.
- Every woman between the ages of 15 to 44 should be fully immunised against tetanus.
- Diarrhoea can kill children by draining too much liquid from the body. So it is essential to give a child with diarrhoea plenty of liquids to drink.
- When a breast-fed child has diarrhoea, it is important to continue breast feeding.
- A child with diarrhoea needs food.
- Trained help is needed if diarrhoea is more serious than usual.
- Diarrhoea can be prevented by breast feeding by immunising all children against measles, by using latrines, by keeping food and water clean, and by washing hands before touching food.
- Families can help prevent pneumonia by making sure that babies are breastfed for at least the first six months of life and that all children are well-nourished and fully immunised.

- A child with cough or cold should be helped to eat and drink plenty of liquids.
- A child with a cough or cold should be kept warm but not hot and should breathe clean non-smoky air.
- Access to safe drinking water and sanitation facilities are the basic necessities.
- You are responsible for your clean environment.
- Illnesses can be prevented by washing hands with soap and water after contact with faces and before handling food.
- Illnesses can be prevented by using latrines.
- Illnesses can be prevented by using clean water.
- Illnesses can be prevented by boiling drinking water if it is not from a safe piped supply.
- Illnesses can be prevented by keeping food clean.
- Illnesses can be prevented by burning or burying household refuse.
- Let the child have access to health care services.
- Health for all by 2000 AD.
- Promote use of low cost technology in health interventions.
- Ensure safe motherhood for child survival.

Right to Protection

- All children are born equal, treat them as equals.
- Each child is an individual, respect him/her.
- Right to equality is a child's birth right.
- Avoid discrimination among children.
- Discrimination among children depletes, human resources.
- Discrimination among children is prohibited.
- Reduce gender disparities among children.
- Give the girl child her rightful share of love, care and protection.

- Girls who are healthy and well-fed during their own childhood and teenage years have fewer problems in pregnancy and childbirth.
- Foeticide and infanticide are evil practice.
- Let the girl child live and prosper.
- Give the girl child equal opportunities for survival and development.
- Communal harmony promotes national integration.
- Child abuse is a sin.
- Child abuse and neglect is a crime.
- Prevent trafficking in children.
- Do not exploit children.
- Immoral traffic among children Is illegal.
- Protect child from all forms of sexual exploitation.
- All parents should tell their children how to avoid getting AIDS.
- Degrading treatment or punishment is inhuman—let the child not suffer.
- Talking, playing and showing love are essential for a child's physical, mental and emotional growth.
- Exemplary punishment to be given to those who involve children in criminal and perverse activities.
- The world must be made safe for democracy.
- Adhere to peace, compassion and cooperation to contain violence.
- So many Gods, so many creeds, so many paths that wind and wind (By Wilock, Eilee Wheeler).

Right to Development

- Children have the right to develop to their full potential.
- Enhance child development by taking care of immediate needs of children.
- Advocate child development programmes.
- Let the best interest of the child be the primary consideration.

- Realise the changing needs, of children.
- Family plays a pivotal role in preserving and transmitting cultural values.
- Strengthen families for well being of the child, individual and society.
- Empower families for the holistic development of the child.
- Reinforce the family's inherent strengths to enable the child to realise its full potential.
- Communication is for development.
- Let the child grow in a healthy family environment.
- Make childhood a happy experience.
- Day care services for children is the need of the hour.
- Separation from parents gives insecurity to children—avoid it.
- Reach the unreached for 'Education for All'.
- Knowledge is power—educate all.
- Each one, teach one.
- Schooling for all is a necessity.
- Primary education is free and compulsory for all children.
- Children have the right to non-discriminatory education.
- Basic education is a developmental necessity.
- Basic education—a right and a necessity.
- Teach child to value and respect his parents.
- Maximise the reach of education.
- Maintain school discipline.
- Handle children with love and care, punishment hampers their development.

Right to Participation

- Respect the views of the child.
- Freedom of expression and thoughts helps in realisation of full potential in a child.

- A child is entitled to privacy.
- Let the child have access to information.
- Books are children's best friend.
- Produce low cost need based communication material.
- Guarantee the child protection against unlawful attack on his honour and reputation.
- Community participation is essential for community education.
- Media advocacy is essential for advancing social development initatives.
- Strengthen child survival and development through selective communication channels and media.
- Social marketing brings attitudinal and behavioural changes.
- Publicise child's rights.

Rights Concerning Effective Implementation of UN Conventions on the Rights of the Child

- Give children's needs a high priority in the allocation of resources.
- Restructure the national budget in favour of human priorities at the government, household and community levels.
- Mobilise human, economic and organisational resources at all levels for the cause of children.
- Mobilise non-traditional resources for actualisation of National Plan of Action.
- Build up stock of human and organisational resources to provide children their basic rights.
- Children first a development priority.
- Prevent human depletion—give children their basic rights.
- National Plan of Action for Children is based on needs, rights and aspirations of children.
- Support implementation of National Plan of Action.

- Implement National Plan of Action to give children their rights.
- Monitor Rights of the Child through social indicators.
- Progress in social indicators is the progress in according the children their rights.

Provisions of Children in Difficult Circumstances in the Convention

- Protect children in difficult circumstances.
- Prevent child labour.
- Let the child develop—do not send him to work.
- Children do not need employment.
- Street childien need care and protection.
- Do not ignore street children.
- Some children cry and die on the streets—save them.
- Ensure survival of abandoned children.
- Abandonment leads to a total deprivation of child's survival, protection and development.
- Help me—find me a family.
- Have one and adopt one.
- Lend your hand in bringing up a child.
- Sponsor a child.
- Promote in-country adoption.
- A disabled child needs special care.
- Provide an environment for survival, protection and development of a disabled child.
- Children who commit crime needs special care.

Give Play Time to the Child

- Play conditions the harmonious development of the child.
- Let the child play.
- Promote toy clubs and banks.
- Involve community in play planning and library.

- Toy Banks in ICDS needs you—Donate generously.
- Give your old toys a home—Donate toys to ICDS.
- Equip the anganwadi—Donate toys.

As Assessment

It is nearly two years since the Government of India ratified the UN Convention on the Rights of the Child (CRC). In a couple of months it has to present a report to the UN on the follow-up action. On the occasion of the Universal Children's Day on November 14, the Department of Women and Child Development released a brochure containing the action taken in respect of the various provisions of the CRC. A rosy picture has been presented in it of the manner in which the Government is fulfilling its commitment to secure the child rights.

The official Report to the UN can be expected to be on the same lines. Yet, before finalising the report, grassroot level workers and experts in child related issues were consulted so their views can be taken into account. The National Consultation, organised by the Indian Council for Child Welfare (ICCW), in collaboration with the UNICEF, brought out sharply the inadequacies in the enforcement of the provisions of the CRC, contrary to the official claims. The CRC, containing 54 Articles, deals with a range of issues concerning child survival and development. At the National Consultation, these were examined under six broad groups: civii rights and freedoms; family environment and alternate care; basic health and welfare; education, leisure and cultural activities; child labour and child abuse; and the role of media. As the main background paper brought out, there is no dearth of legislative and constitutional provisions to facilitate the child's right to fulfilment of its basic needs. But, the enforcement of these has been far from satisfactory.

The very definition of child has become an issue. Different laws have different age limits to define a child, ranging from 12 to 18. The CRC defines a child as under the age of

18. Acceptance of this age limit has implications for the various programmes and budgetary provisions made for child-related schemes which are based on the existing definition. However, the Government is reviewing the various legislative measures and is considering the adoption of the definition in the CRC. If the CRC definition is accepted, the child population will be much higher than the current figure of 300 million which itself represents over one-third of the total population.

This is a significant proportion and cannot be ignored. It was against this background that the participants deliberated on how to remove the existing lacunae and ensure the child its rights. Access to education is one of the basic rights of the child, which the Government is committed to provide. The commitment precedes the CRC. But very little has been achieved towards fulfilling this. In recent years, there have been fresh initiatives in this direction. But the group of educationists and grassroot-level workers, who participated in the National Consultation, felt that these initiatives implied dilution, rather than fulfilment of the constitutional obligations. The Government has, contrary to the constitutional obligation of providing free and compulsory education to all children upto 14 years of age, shifted its focus to universalisation of primary education (UPE) covering children in the age group of 6-11 only. The children in the age group 11-14 have been left in the lurch.

Of course, there is a programme for universalisation of elementary education (UEE) to cover the age group 11-14 but the priority is for the UPE. The launching of the non-formal education and the national literacy mission programmes is also viewed as a shift away from the constitutional obligation of providing integrated education from the early childhood to 14 years of age, and away from the commitment under the CRC to provide education up to 18 years. Even the target dates for achieving the UPE and UEE goals have been put off by five years. The UPE target is now set at 1995 and the UEE target at 2000. Child labour is another issue on which the Government approach has been found wanting. A point was made that the Government policies were based on flawed assumptions and faulty appreciation of the field situation. Hence, it was felt that no significant change could be made in the child labour situation unless the basic premises were abandoned and a different view was adopted.

The two premises to be questioned are:

(a) that child labour is a harsh reality attributable to poverty of the parents; and

(b) that there is a distinction between exploitative child labour and child labour under family environment.

On the basis of this, the extent of child labour has been measured and the number of working children estimated at 18.17 million in 1990 and 20.15 million in the year 2000. Non-official estimates put the figure at as high a level as 144 million. According to the expert group at the National Consultation, all children who do not go to school should be considered as working children. On this basis the figure would be 90 million. Viewed in this light, the inadequacy of Government proposal to cover two million working children over a period of six years under its programme of elimination of child labour by 2000 AD is, blatant. Even if the official estimate of 20 million is accepted, the coverage is just 10 per cent and, if the estimate of 90 million is taken, it works out to a mere two per cent. What dent can this be expected to make on the problem?

Moreover, the very content of Government policy is such as to encourage child labour rather than discourage it. For instance, the nonformal education scheme in essence provides for coexistence of child labour and child education. To that extent, the child labour is allowed to continue without hindrance. The experts lament that, despite criticism of its politics, the Government has not chosen to make amends and, instead, has announced a further plan of action to eliminate child labour on the same lines.

The alternative suggested is that the Government should, in accordance with Article 32 of the CRC, regard all non-school-going children as working children, instead of only those in hazardous occupations, raise allocations for education to six per cent of GNP, rely on formal education rather than on non-formal education, improve the formal education to make it more attractive and make it compulsory on the part of the parents to send their children to school and not to work.

In regard to the child's right to basic health services, too, the official claims have been received with a discount. True,' the Infant Mortality Rate (IMR) has been reduced over the years. But, this is at the level of national average. There are still pockets with high IMR. While in Kerala it is as low as 17 per thousand live births, in Orissa it is as high as 114 as compared to the national average of 74. Statistically the infrastructure for primary health care delivery has expanded vastly but the ground reality is that there are no doctors or medicines in many PHCs. The official claims of achievement in the provision of basic health services have been dismissed as a "mere number game."

9

UN Convention on the Rights of the Child

It is pertinent to note that if 'rights' are legally recognised interests, the obvious answer is that legal institutions and legal sanctions should ensure the enjoyment of rights. The problem with children, however, is that they by themselves are unable to seek protection under law because of their tender age and inability to command resources for accessing legal services. The legal system envisages that people are aware of their rights and, in case of breach, they are capable of asserting them through proper legal channels. Both these preconditions do not exist in reality in case of children. As such, the legal system ought to have alternate mechanisms to act on behalf of the child if it were to protect child rights when threatened. It is here that the role of the State and the society including voluntary organisations assumes crucial importance for. delivery of justice to the child.

It must be remembered that it took several centuries for legal systems everywhere to recognise an independent personality for the child capable of having rights even against the parents and guardians. However, since the establishment of the United Nations and the adoption of the Universal Declaration of Human Rights there has been several changes in the laws of different countries including India conceding preferential rights and entitlements to children. The General Assembly of the United Nations did adopt as early as 1959 a Declaration on the Rights of the Child laying down certain basic principles in dealing with children. The world body followed it up in 1989 with a full-fledged Convention on the Rights of the Child with a built-in mechanism for its enforcement by participating countries.

With the ratification of the Convention by most governments, we have today a body of international law which can be invoked to protect the rights of children by their respective governments. The Convention reiterates the principle of equality and non-discrimination (Article 2) promises the survival, protection and development of children (Articles 23 to 39), and obligates the state to ensure protection of all the rights in the best interests of the child (Articles 3 and 4). It is undoubtedly the most comprehensive document on child rights which at least for the rest of the present century will guide the destiny of millions of children around the globe on a firm legal footing. If one were to examine closely the provisions of the Convention and interpret them from a positive law angle, one cannot avoid implementational problems particularly when the Governments aw not favourably disposed. For one thing, the rights excepting in certain cases, are broadly worded, vaguely defined and made dependent on the resources available to State parties.

There is very little commitment reflected in the Convention to support the realisation of the rights of children everywhere in the foreseeable future which requires resources of the size few countries can afford at present If environment protection can command a fund of its own, one fails to understand why it should not be there for child rights as well. UNICEF as it is structured and funded, cannot be expected to fulfil the need for greater resources in this regard unless a strategic shift takes place in the UN system and in the policies of funding agencies like World Bank and IMF.

Data presented at the International Conference on Nutrition (Rome, December 1992) showed that the number of malnourished children worldwide increased from 166 million in 197S to 188 million in 1990. A contributing factor for this unfortunate situation lies in the fact that 82.4 per cent of the annual global income went to less than 20 per cent of the human population, the poorest 20 per cent receiving less than 2 per cent of the annual world income (UNDP's Human Development Report, 1992). The future appears to be grim for India's malnourished and neglected children at least in the short run, if the experience of other developing countries which have undergone structural adjustment programmes is to be believed.

India's level of spending on social services is already low and it may further deteriorate under the liberalisation programme in which case the Convention is of little use in improving the status of the child. There are, of course, certain rights which either require only very little finances or are to be honoured irrespective of resources available. There is thus a vast area of child rights awaiting vigorous enforcement at the instance of governments and voluntary organisations. They include civil rights and freedoms (Articles 6 to 22), protection from torture and other cruel or inhuman treatment (Articles 37 and 39) and the child's rights with regard to due process of law and special procedures in criminal proceedings (Article 40).

In so far as children are ignorant of their rights and are not in a position to assert them through legal channels, there is a great responsibility on State institutions and more particularly on the judiciary to be pro-active, non-technical and accommodative of the best interests of the child in dealing with situations involving them. The voluntary organisations and the media have unlimited scope for rendering useful services at different levels in the delivery of justice to the child. It is stated that by ratifying the Convention, the child rights agenda ought to receive the first call of all governmental resources. If that is so, there is tremendous scope for NGOs, political parties, professional associations, media groups and opinion leaders to lobby for better allocation of resources to child welfare and to demand accountability from governmental institutions towards honouring the rights of children in governmental programmes. The Convention expects the Government to review the existing laws relating to children and bring them in line with its provisions.

The process has already begun and voluntary organisations can help in not only speeding up the process but also in structuring the reform in a manner appropriate to the best interests of children. Publicising the Convention and making child rights known at least to all adults including parents is an obligation not only of Government but of every service organisation. This is a priority task which may not need huge monetary resources if imaginative steps are adopted. If awareness grows, implementation will be facilitated and monitoring becomes easy. The Convention will have fulfilled some purpose if the legislative agenda and the publicity exercises are attended to on a priority basis.

So long as India remains a democracy governed by rule of law, the above strategies are bound to generate positive results even when resources do not permit revolutionary changes on the status of children. The recent experiences in evolving the country report on the implementation of the Convention after a series of consultation between Government agencies and NGOs, is a welcome development pregnant with possibilities for better enforcement of child rights.

Constitutional rights are known in India as Fundamental Rights. They are available to all citizens including children. They cannot be taken away by legislation or governmental action excepting within constitutionally permitted limits. There are special, expeditions and cheap procedures to seek protection of fundamental rights either from the High Court or from the Supreme Court directly. These procedures have been further liberalised by courts through what is known as public interest litigation enlarging the scope for easy access to judicial remedies. Courts are empowered even to strike down parliamentary legislations which infringe fundamental rights. Courts do in fact liberally interpret the constitutional rights invoking International Human Rights Instruments as well as the Directive Principles of State Policy contained in Part IV of the Constitution.

Thus, today an Indian citizen enjoys far too many rights than what was assumed forty years ago when the original Constitution was adopted. It is therefore imperative for every one including children to understand the Indian Constitution and its pivotal role in organising governance in a manner that protects human rights of everyone particularly of the weak and vulnerable sections of society. In this context it will be appropriate to examine the nature and scope of rights and freedoms which the constitution provides for. The constitutional treatment of rights is two-fold. On the one hand, certain rights and freedoms which are mostly of civil, political and cultural in nature are secured as fundamental rights enforceable against the State through constitutionally guaranteed remedies including writs. Though these rights are couched in negative language, they are capable of assuming the character of positive rights enforceable through affirmative action on the part of the State and its agencies.

These set of rights include (a) right to equality (Articles 14-18); (b) right to freedom (Articles 19-22), (c) right against exploitation (Articles 23-24); (d) right to freedom of religion (Articles 25-28); (e) cultural and educational rights (Articles 29-30); and (f) right to constitutional remedies (Article 32).

On the other hand certain rights which are largely social and economic in nature have been listed as "Directive Principles of State Policy" and have been made not justiciable through courts of law. In this regard, the language used is positive inasmuch as they are made *fundamental in governance* and are considered a necessary supplement to fundamental rights in achieving a Welfare State. In fact, Parliament is empowered even to amend the fundamental rights for implementing the Directives, so long as the amendment does not touch the basic features of the Constitution. In other words, the now accepted norms of Indian Constitutional law declare that legislations intended to implement the rights involved in the Directives have to be held constitutionally valid as far as possible.

This recognition of "indivisibility" and "inviolability" of rights included in the Directive Principles (Social and Economic Rights) and the harmonious interpretation of other provisions in the Constitution to advance the content of the rights in Part IV (Directive Principles of State Policy) is a major accomplishment in human rights protection under Indian constitutional law. The Indian Constitution does command the State "to respect international law and treaty obligations" (Article 51) and the court, as part of the State, has been consistently upholding human rights standards by invoking the appropriate Conventions and Declarations. A quick look at the Constitution presents an impressive list of rights for children, some of them not even available to adults. While the right of "equality before the law and the equal protection of the laws" is available to any person (including children), Article 15 (3) empowers the State to have special laws for children only intended to enable them to enjoy the fruits of the equality guarantee.

Thus the child labour prohibition laws, the Juvenile Justice Act and the Children's Act, though are a class legislations, are not only constitutionally permissible but also constitutionally mandated. Secondly, taking note of the then prevailing exploitative practice of trafficking in human beings, particularly children, Article 23 of the Constitution has put a total ban on forced labour and made such practices punishable under law. Showing its concern for the tender age of children, the Constitution in Article 24 has prohibited the employment of children below the age of 14 years in any factory, mine or in any other hazardous activity. Besides, all the seven sets of fundamental rights in the Constitution are available to children with as much authority and accessibility as that with adult citizens. There is a large body of case law which developed the application of these rights to children and have expanded their scope in order to make them meaningful to children. This was accomplished by the judiciary by using the technique of harmonious construction of Part-Ill (Fundamental Rights) and Part IV (Directive principles of State Policy) of the Constitution side by side with India's Treaty obligations and obligations under the UN Conventions on the rights of the Child. In this regard, the performance of the Indian judiciary stands out as a signal contribution for the universal development of human rights generally and that of child rights in particular.

What are those provisions in Part IV of the Indian Constitution which enabled the judiciary to promote the jurisprudence of child rights? They are briefly Articles 39 (e), (f), 42,45 and 47 which are reproduced here.

Article 39

The State shall, in particular, direct its policy towards securing:

(e) that the health and strength of workers and the tender age of children are not abused and that citizens are not involved by economic necessity to enter avocations unsuited to their age or strength;

(f) that children are given opportunities and facilities to develop in a healthy manner and in conditions of freedom and dignity and that childhood and youth are protected against exploitation and against moral and material abandonment.

Article 42

The State shall make provision for securing just and humane conditions of work and for maternity relief.

Article 45

The State shall endeavour to provide, within a period often years from the commencement of this constitution, for free and compulsory education for all children until they complete the age of fourteen years.

Article 47

The State shall regard the raising of the level of nutrition and the standard of living of its people and the improvement of public health as among its primary duties and, in particular, the State shall endeavour to bring about prohibition of the consumption except for medicinal purposes of intoxicating drinks and of drugs which are injurious to health.

Very few persons counted these as positive rights partly because they are expressly made non- enforceable through courts (Article 37) and partly because they are meant to be only progressively realisable subject to the economic resources of the State. At the same time Article 37 has stipulated in unequivocal terms that "these principles are fundamental in the governance of the country and it shall be the duty of the State to apply these principles in making laws." The Supreme Court therefore, as part of the system of governance of the country, is justified in invoking these principles in interpretation of statutes and in implementation of fundamental human rights. Thus the failure on the part of the State to provide free, compulsory education for all children *within a period of ten years from the commencement of the Constitution* has compelled the Supreme Court to declare right to education as part of the fundamental right to personal liberty (Article 21) as without education, life cannot be lived with dignity. This tendency to read the rights contained in Directive Principles (Part IV) into the fundamental rights in Part III of the Constitution is a welcome contribution of the judicial wing of the government towards progressive realisation of the rights under U.N. Declarations and Conventions. In fact, in most of the opinions of the Supreme Court, a major support for interpretative logic has been the principles contained in the U.N. Convention to which India is a signatory. This has led to a new approach in legal engineering often referred to as the domestication of the international human rights norms, standards and procedures, through the judicial process. In this approach one group of citizens who are the clear winners is the hitherto neglected section of children who have not been able to assert their rights through courts.

However, in one sphere the judiciary has failed in its responsibilities towards children. This is in respect of processual justice under the Juvenile Justice Act. There are reports that in some States the juvenile courts do not have presiding officers and if they do have, they are people not necessarily the best suitable persons available. The charge of inordinate delay in disposal of cases is indeed a fact in many juvenile court proceedings. Unimaginative disposition of juvenile delinquents and lack of infrastructural support have made the whole system counterproductive, authoritarian and inimical to child rights. There has not been a proper study of the strengths and weaknesses of the system and it appears to be not a priority item to the higher judiciary which controls the functioning of juvenile courts. If the judiciary really cares for the rights of the child, it is imperative that it bestows greater attention to juvenile courts, its personnel, its procedures and infrastructural facilities available to it. Perhaps the system of public participation provided for by law is required the most in the juvenile justice area than in any other field and attitude and approaches of the judiciary will have to change to attract this necessary input in dealing with children in conflict with the laws.

Gaps and Inadequacies in Legislation Concerning Children

It is neither necessary nor feasible to examine the large number of laws already available in the Statute Book giving protection to the rights of the child. They make India one of the leading countries in legislative protection of children though in implementation of the laws, the country is notoriously soft and backward. Nevertheless, after the ratification of the Convention, and in the spirit of the

Constitutional mandates, there are still some crucial areas left where legislation is scanty and inadequate to protect all the rights of children. Some of these instances may be examined in the context of the obligations under the Convention.

One such instance is the subject of adoption which indeed is a boon for thousands of underprivileged children in India. Despite repeated attempts by private members to have an adoption law projecting "the best interests of the child", the Centre did not enact one because of minor opposition from some quarters. A Uniform Civil Code incorporating the equality rights of women and children though part of the Directives of the Constitution, did not attract the attention of Government thereby jeopardising the rights and entitlements otherwise available to women and children. Today the darkest chapter of Indian law is in the area of family relations which are still governed by a bewildering variety of customary, religion-based and feudalistic personal laws applicable to persons depending on their birth into one or other religion. Again there are gaps in laws relating to health, education, housing and employment where compromises have been made in different degrees to the pre-existing practices derogatory to the dignity of childhood and the human rights-based principles of the Constitution and the Convention. Absence of compulsory primary education law and continued legitimacy to employment of children in certain cases are instances in point. It is doubtful whether children have any housing right at all under the existing laws. On child prostitution, on the status of disabled children and on the neglect of the so-called "street children" there are serious inadequacies and shortcomings in the law when compared to the principles and standards of the Convention.

Voluntary Agencies and Implementation of Rights of the Child

It is acknowledged on all sides that a little better implementation of the existing laws will secure greater protection to the child than seeking more laws and more governmental bureaucracies for exercising power under them.

In considering the status of child rights in relation to the Convention, the Constitution and the laws and in the context of social action, one may suggest the following issues for priority consideration.

1. Can the review and revision of laws which is an obligation under the Convention, be referred to a Child Law Commission involving besides law persons, activists, academicians, behavioural scientists, medical and media personnel ? Should not such a Commission be asked to simplify and consolidate all child care related laws in one Code incorporating the substantive and procedural aspects for easy access to everybody concerned including children?

   Can it be declared as a role of practice that all future legislative bills must necessarily have an explanatory memorandum about the possible impact of the proposed legislation on the child population?
2. Can a body parallel to the Supreme Court be created by Parliament under Article 32(3) of the Constitution to oversee and adjudicate enforcement of child rights at the national level?
3. How to organise efficiently training programmes on child rights for the civil services, judiciary, police, media persons, parents and community leaders? Can it be organised in such a way to take off with the setting up of Panchayats and local bodies?
4. How to generate adequate financial and technical resources to realise child rights in the foreseeable future? What is the obligation of international institutions and NGOs in this regard.
5. What is the best arrangement to organise appropriate legal services to children in need of such services? The existing legal aid programme

be adapted to the unique needs of children particularly when adults are in the other side?

6. Are the responses of law to child rape, child prostitution and child abuse adequate? If not, how can law be made purposeful to advance the *best interests of the child?*
7. Can a uniform adoption law be brought into force to uphold the interests of children ? What are the strategies and approaches therefore?
8. Given the cultural traits and traditions of the Indian situation what are the special provisions the legal system should have to protect the girl child from discriminatory treatment.
9. As "education" is admittedly the Master Right to realise all other rights, how to go about realising this right to all children in the near future? What legal framework is appropriate to facilitate the enjoyment of the right to education by all children ?
10. How can the justice institutions including the juvenile courts be re-organised to cater to the needs of children expeditiously and effectively in the best interests of children?

India's accession to the UN Convention on the Rights of the Child in December 1992, was an important step towards initiating accountability and reinforcement of the Indian State's commitment towards the survival, protection and development of its children. As outlined by Article 43 of the Convention, all State parties who are signatories to the Convention are obliged to present a report to the Committee on the Rights of the Child for "the purpose of examining the progress made by State Parties in achieving the realisation of the obligations undertaken in the present Convention." The Indian Government is expected to submit its report on the measures they have adopted which give effect to the rights outlined in the Convention to the Committee on the

Rights of the Child in Geneva in January 1995. This report is of crucial importance as it will reflect the situation of children in India vis-a-vis their access to rights that the international community has recognised as being crucial to their development and that assures a childhood in conditions of freedom and dignity.

It has been 20 long years since the National Policy on Children was formulated, wherein the Government of India promised its children a generous deal. The Constitution of India recorded a firm commitment that the basic needs of the child would be met. However the balance sheet of the Government is a long list of failures more than a record of achievements that benefit the Indian children.

In the area of IMR (Infant Mortality Rate)—there has been a reduction from a level of 129 deaths per 1000 live births in 1971 to 79 in 1992. Yet, of the 25 million children born every year in India, two million die before reaching the age of one, primarily due to malnutrition and preventable diseases. Similarly, in the field of education, the Government had visualised higher percentage of children finishing primary school. The enrolment figures of children in schools have gone up from 19.2 millions in 1951 to 105 million in 1993, yet this increase covers only 33% of the total child population. The rate of retention in schools continues to be low i.e. only 50% of enrolled primary school children reach class V. There is still a long way to go—not only in getting children to schools but also in retaining them, and reducing the dropout and stagnation rates. In fact, half of the children in the age group 5-14 years are illiterate. Nearly 55% of children do not attend schools. Quality of education and the competencies developed by those who complete primary school is another area of concern.

The issue of child labour has been researched, debated and discussed for the past decade and a half. Solutions to eradicate child labour have been put forth and there are numerous programmes being run by NGOs and Government, but we still have child labourers with their numbers increasing. It is estimated that there are anywhere between 44 million (1983, ORG study, Baroda) to 110 million (Balai Data Bank, Manila, Philippines) children in our country who sweat it out working in most oppressive and exploitative conditions. Children have waited eight long years for the implementation of the Child Labour (Prohibition and Regulation) Act, 1986. It is still a dream and hope.

Child abuse and neglect is an area with dearth of empirical data on its magnitude, nature, causes and effects of forms of child abuse and neglect primarily due to absence of any clear definition of "child abuse" and "child neglect". Most of the data is in the form of journalistic pieces and case studies or isolated micro studies. The worst form of abuse is sexual abuse. A study done in 1985 by Tata Institute of Social Sciences shows that one out of every 3 girls and one out of every 10 boys were sexually abused in their childhood. Further, the study claims that most sexual abuse takes place within the family itself. Most of these cases came to light because the mothers took the courage to come out in the open and lodge complaints at the police stations. All types of sexual abuse are punishable under IPC.

Child prostitution is the worst form of abuse and exploitation. An NGO estimates that the number of child sex workers in India is around 4,00,000. It is also a matter of concern that despite the ban, about 5000 girl children are annually inducted in the Devadasi practice in at least eight districts of Maharashtra and Karnataka. Although Immoral Trafficking (Prevention) Act 1956 prohibits child prostitution, the sexual exploitation of children still continues. In fact, there are strong fears of child prostitution increasing as clients seem to be preferring children to adult sex workers with fear of HIV- AIDS infection.

There are a number of reported cases of sale and

trafficking of minors, however there is no documentation on the modalities, routes and factors contributing to child trafficking. Much less is known or documented on what happens to the victims or the violator.

Similarly if we were to assess several other aspects of child welfare and development, there is evidence that much more needs to be done.

Two years ago i.e. 1992, Government of India formulated a National Plan of Action for children setting a target of few immediate goals to improve the lives of children in our country. They are as follows:

Education

- Access to enrolment in primary education for at least 80% of boys and 75% of girls by 1995.
- Completion of Primary Education by at least 50% of girls as well as boys by 1995.
- Reduction of adult and adolescent illiteracy from 1990 levels fey 25% by 1995.

Health

- Universal use of oral rehydration therapy for home based treatment of diarrhoea and universal access to oral dehydration salts by 1996.
- Access to family planning services in order to increase contraceptive prevalence levels by 50% from the current rate by the year 2000.

Nutrition

- Universal access to iodised salt by 1995.

Child Labour

- Progressive and accelerated elimination of child labour.

Safe Environment

- Access to safe drinking water to not less than 3/4th of both rural/urban population by 1996 and universal access by the year 2000.
- Double current levels of access to sanitary means of excreta disposal by 1996.

APPENDICES

APPENDIX — I

DECLARATION OF THE RIGHTS OF THE CHILD

Passed by the General Assembly of the United Nations in 1959.

Preamble

Whereas the people of the United Nations have, in the Charter, reaffirmed their faith in fundamental human rights, and in the dignity and worth of the human person, and have determined to promote social progress and better standards of life in larger freedom,

Whereas the United Nations has, in the Universal Declaration of Human Rights, proclaimed that everyone is entitled to all the rights and freedoms set forth therein, without distinction of any kind, such as race, colour, sex, language, religion, political or other opinion national or social origin, property, birth or other status,

Whereas the child, by reason of his physical and mental immaturity, needs special safeguards and care, including appropriate legal protection, before as well as after birth,

Whereas the need for such special safeguards has been stated in the Geneva Declaration of the Rights of the Child of 1924, and recognised in the Universal Declaration of Human Rights and in the statutes of specialised agencies and international organisations concerned with the welfare of children,

Whereas mankind owes to the child the best it has to give, *Now therefore,*

The General Assembly

Proclaims this Declaration of the Rights of the Child to the end that he may have a happy childhood and enjoy for his own good and for the good of society the rights and freedoms herein set forth, and calls upon parents, upon men and women as Individuals and upon voluntary organisations, local authorities and national Governments to recognise these rights and strive for their observance by legislative and other measures progressively taken in accordance with the following principles:

Principle 1

The child shall enjoy all the rights set forth in this Declaration. All children, without any exception whatsoever, shall be entitled to these rights, without distinction or discrimination on account of race, colour, sex, language, religion, political or other opinion, national or social origin, property, birth or other status, whether of himself or of his family.

Principle 2

The child shall enjoy special protection, and shall be given opportunities and facilities, by law and by other means, to enable him to develop physically, mentally, morally, spiritually and socially in a healthy and normal manner and in conditions of freedom and dignity. In the enactment of laws for this purpose the best interests of the child shall be the paramount consideration.

Principle 3

The child shall be entitled from his birth to a name and a nationality.

Principle 4

The child shall enjoy the benefits of social security. He

shall be entitled to grow and develop in health; to this end special care end protection shall be provided both to him and to his mother, including adequate pre-natal and post-natal care. The child shall have the right to adequate nutrition, housing, recreation and medical services.

Principle 5

The child who is physically, mentally or socially handicapped shall be given the special treatment, education and care required by his particular condition.

Principle 6

The child, for the full harmonious development of his personality, needs love and understanding. He shall, wherever possible, grow up in the care and under the responsibility of his parents, and in any case in an atmosphere of affection and of moral and material security; a child offender years shall not, save in exceptional circumstances, be separated from his mother. Society and the public authorities shall have the duty to extend particular care to children without a family and to those without adequate means of support. Payment of State and other assistance towards the maintenance of children of large families is desirable.

Principle 7

The child is entitled to receive education, which shall be free and compulsory, at least in the elementary stages. He shall be given an education which will promote his general culture, and enable him on a basis of equal opportunity to develop his abilities, his individual judgement, and his sense of moral and social responsibility, and to become a useful member of society.

The best interests of the child shall be the guiding principle of those responsible for his education and guidance; that responsibility lies in the first place with his parents.

The child shall have full opportunity for play and recreation, which should be directed to the same purposes as education; society and the public authorities shall endeavour to promote the enjoyment of this right.

Principle 8

The child shall in all circumstances be among the first to receive protection and relief.

Principle 9

The child shall be protected against all forms of neglect, cruelty and exploitation. He shall not be the subject of traffic, in any form.

The child shall not be admitted to employment before an appropriate minimum age; he shall in no case be caused or permitted to engage in any occupation or employment which would prejudice his health or education, or interfere with his physical, mental or moral development.

Principle 10

The child shall be protected from practices which may foster racial, religious and any other form of discrimination. He shall be brought up in a spirit of understanding, tolerance, friendship among peoples, peace and universal brotherhood and in full consciousness that his energy, and talents should be devoted to the service of his fellow men.

APPENDIX —II

THE CONVENTION ON THE RIGHTS, OF THE CHILD

The Convention on the Rights of the Child is the most widely ratified human rights convention history. About 96 per cent of the world's children live in States that have recognised their rights a are legally obliged to fulfil them. Only six countries have still to ratify the Convention: the Co Islands, Oman, Somalia, Switzerland, the United Arab Emirates and the United States.

The Convention on the Rights of the Child is helping to shape legal codes and school curricula, well as public and official attitudes about children and their rights.

In Rwanda, conditions are improving for adolescents facing prosecution on charges stemming from the 1994 genocide and war. By October 1996, about 200 had been moved from overcrowded ad prisons to UNICEF-supported rehabilitation centres. Training on child rights is being provided to prison, judicial, police and military-police officials.

In Ecuador, a campaign headed by the First Lady is promoting awareness of every child's right a name and a nationality. The goal is to register some 400,000 children under the age of 12 who have no identity documents.

In Mwanza, Tanzania's second largest city, an advocacy group called Kuleana has helped sensitise police, city magistrates and lawyers about children's rights. This has led to less violence against children living on the street and to fewer of them receiving, criminal convictions.

Romania has launched a pilot project in Bucharest to give juvenile offenders who would normally be sent to reform school the opportunity to remain with their families, receive counselling by social workers and carry out community service work in public institutions.

In San Ramon, Nicaragua, the Movimiento Infantil (Children's Movement) organised protests against Physical punishment by teachers—and ran workshops sensitising teachers about child rights. The group also successfully lobbied the municipal government to improve public sanitation.

BIBLIOGRAPHY

Ahearn, D., Holzer, B. with Andrews, L. (2000, 2007) *Children's Rights Law: A Career Guide*. Harvard Law School. Retrieved 2/23/08.

Ariès, Philippe. Centuries of Childhood: A Social History of Family Life. New York: Alfred A. Knopf, 1962.

Arts, K, Popvoski, V, *et al.* (2006) *International Criminal Accountability and the Rights of Children*. "From Peace to Justice Series". London: Cambridge University Press.

Bandman, B. (1999) *Children's Right to Freedom, Care, and Enlightenment*. Routledge. p 67.

Bihar State Council for Child Welfare. The Convention on the Rights of the Child: A State Level Workshop for NGOs in Bihar, 1994, Oct. 4-6, Walmi. Patna, the author, 1994.

Boas, George. The Cult of Childhood. London: Warburg, 1966.

Brennan, S. and Noggle, R. (1997) "The Moral Status of Children: Children's Rights, Parent's Rights, and Family Justice," *Social Theory and Practice. 23.*

Brown, Marilyn R., ed. Picturing Children: Constructions of Childhood between Rousseau and Freud. Aldershot: Ashgate, 2002.

Brownlie, J. and Anderson, S. (2006) "'Beyond Anti-Smacking': Rethinking parent–child relations," *Childhood. 13*(4) p 479-498.

Buckingham, David. After the Death of Childhood: Growing Up in the Age of Electronic Media. Blackwell Publishers, 2000.

Bunge, Marcia J., ed. The Child in Christian Thought. Grand Rapids, MI: William B. Eerdmans Publishing Co., 2001.

Calkins, C.F. (1972) "Reviewed Work: Children's Rights: Toward the Liberation of the Child by Paul Adams", *Peabody Journal of Education. 49*(4). p. 327.

Calvert, Karin. Children in the House: The Material Culture of Early Childhood, 1600-1900. Boston: Northeastern University Press, 1992.

Cannella, Gaile and Joe L. Kincheloe. "Kidworld: Childhood Studies, Global Perspectives, and Education". New York: Peter Lang, 2002.

Cleverley, John and D.C. Phillips. Visions of Childhood: Influential Models from Locke to Spock. New York: Teachers College, 1986.

Covell, K. and Howe, R.B. (2001) *The Challenge of Children's Rights for Canada. Wilfrid Laurier University Press.* p 158.

Cunningham, Hugh. Children and Childhood in Western Society since 1500. London: Longman, 1995.

Cunnington, Phillis and Anne Buck. Children's Costume in England: 1300 to 1900. New York: Barnes & Noble, 1965.

Cutting, E. (1999) "Giving Parents a Voice: A Children's Rights Issue," *Rightlines. 2* ERIC #ED428855.

DeLamater, J.D. (2003) *Handbook of Social Psychology.* Springer. p 150.

deMause, Lloyde, ed. The History of Childhood. London: Souvenir Press, 1976.

Dua, M. R. and Gupta V. S. (Ed.) *Media and Development: Theories in Communication and Extension.* New Delhi, Har-Anand Publication, 1994.

Franklin, B. (2001) *The new handbook of children's rights: comparative policy and practice.* Routledge. p 19.

Freeman, M. (2000) "The Future of Children's Rights," *Children & Society. 14*(4) p 277-93.

Higonnet, Anne. Pictures of Innocence: The History and Crisis of Ideal Childhood. London: Thames and Hudson Ltd., 1998.

Immel, Andrea and Michael Witmore, eds. Childhood and Children's Books in Early Modern Europe, 1550-1800. New York: Routledge, 2006.

Jenks, C. (1996) "Conceptual limitations," *Childhood.* New York: Routledge. p 43.

Joseph M. Hawes, *The Children's Rights Movement: A History of Advocacy and Protection* (Boston: Twayne Publishers, 1991).

Kaslow, FW (1990) Children who sue parents: A new form of family homicide? *Journal of Marital and Family Therapy. 16*(2) p 151–163.

Kincaid, James R. *Child-Loving: The Erotic Child and Victorian Culture.* New York: Routledge, 1992.

Knörr, Jacqueline, ed. *Childhood and Migration. From Experience to Agency.* Bielefeld: Transcript, 2005.

Lansdown, G. (1994). "Children's rights," in B. Mayall (ed.) *Children's childhood: Observed and experienced.* London: The Falmer Press. (p 33-34).

Lansdown, G. "Children's welfare and children's rights," in Hendrick, H. (2005) *Child Welfare And Social Policy: An Essential Reader.* The Policy Press. p. 117

League for Children's Rights Individual UPR Submission: Germany. February 2009. Submitted by Bündnis RECHTE für KINDER e.V. and supported by President of the INGO Conference of the Council of Europe, Annelise Oeschger. Retrieved December 27, 2011.

Maheshwar, Madan Lal. Role of Media in Creating Awareness for Children Rights in *A Report of the Workshop on the Rights of the Child,* 1994 August 30-31, Jabalpur, M.P.

Mahila Dakshata Samithi. *State Level Workshop on Rights of the Child: A Report.* 1994, Sep. *27-28,* Hyderabad. Hyderabad.

Mangold, S.V. (2002) "Transgressing the Border Between Protection and Empowerment for Domestic Violence Victims and Older Children: Empowerment as Protection in the Foster Care System," New England School of Law. Retrieved 4/3/08.

Mason, M.A. (2005) "The U.S. and the international children's rights crusade: leader or laggard?" *Journal of Social History*. Summer.

Mathur, Kanwar B. *Communication for Devëlopment and Social Change*. New Delhi, Allied Publishers, 1994.

Melchiorre, A. (2004) At What Age?...are school-children employed, married and taken to court?

Müller, Anja, ed. *Fashioning Childhood in the Eighteenth Century:* Age and Identity. Burlington, VT: Ashgate, 2006.

O'Malley, Andrew. The Making of the Modern Child: Children's Literature and Childhood in the Late Eighteenth Century. London: Routledge, 2003.

Pinchbeck, Ivy and Margaret Hewitt. Children in English Society. 2 vols. London: Routledge, 1969.

Pollock, Linda A. Forgotten Children: Parent-child relations from 1500 to 1900. Cambridge: Cambridge University Press, 1983.

Rachel K. Jones and April Brayfield, Life's greatest joy?: European attitudes toward the centrality of children. *Social Forces,* Vol. 75, No. 4, Jun 1997. 1,239-69 pp. Chapel Hill, North Carolina.

Rodham, H. (1973). "Children Under the Law". Harvard Educational Review 43: 487–514.

Rojas Flores, J. (2007) "The rights of the child in Chile: an historical view, 1910-1930", *Historia. 40*.3. Special issue.

Sommerville, C. John. *The Discovery of Childhood in Puritan England.* Athens: University of Georgia Press, 1992.

Starr, RH (1975) *Children's Rights: Countering the Opposition.* Paper presented at the 83rd Annual Meeting of the

American Psychological Association in Chicago, Illinois, Aug. 30-Sept. 3, 1975. ERIC ID# ED121416.

Steinberg, Shirley R. and Joe L. Kincheloe. *Kinderculture: The Corporate Construction of Childhood.* Westview Press Inc., 2004.

Stone, Lawrence. The Family, Sex and Marriage in England 1500-1800. New York: Harper and Row, 1979.

Thorne, B. (1987) "Re-Visioning Women and Social Change: Where Are the Children?" *Gender & Society 1*(1) p. 85–109.

UNICEF. *Rights of the Child.* New Delhi, the author, *1992.*

UNICEF. *The State of the World's Children,* 1995. New Delhi, the author, 1994.

W.J. Rorabaugh, Donald T. Critchlow, Paula C. Baker (2004). "America's promise: a concise history of the United States". Rowman & Littlefield. p. 47.

Windahl, Sven *et al. Using Communication Theory: An Introduction to Planned Communication.* London, Sage Publications 1992.

Zornado, Joseph L. *Inventing the Child: Culture, Ideology, and the Story of Childhood.* New York: Garland, 2001.

INDEX